Anatoly S. Rozhkov
Tatyana A. Mikhailova

The Effect of Fluorine-Containing Emissions on Conifers

Translated by L. Kashhenko

With 28 Figures and 26 Tables

Springer-Verlag
Berlin Heidelberg New York
London Paris Tokyo
Hong Kong Barcelona
Budapest

Professor Dr. Anatoly S. Rozhkov
Dr. Tatyana A. Mikhailova

Siberian Institute
of Plant Physiology and Biochemistry
Siberian Branch
of the Russian Academy of Sciences
P.O. Box 1243
Irkutsk-33, 664 033
Russia

Library of Congress Cataloging-in-Publication Data
Rozhkov, A. S. (Anatoliĭ Sergeevich) [Deĭstvie ftorsoderzhashchikh ėmissiĭ na khvoĭnye derev'i͡a. English] The effect of fluorine-containing emissions on conifers/Anatoly S. Rozhkov, Tatyana A. Mikhailova. p. cm. Includes bibliographical references and index.
ISBN-13: 978-3-642-77052-4 e-ISBN-13: 978-3-642-77050-0
DOI: 10.1007/ 978-3-642-77050-0
1. Conifers – Effect of fluorine on. 2. Conifers – Wounds and injuries. 3. Conifers – Effect of air pollution on. 4. Fluorine – Environmental aspects. 5. Aluminum industry and trade – Environmental aspects – Russia (Federation) – Siberia, Eastern. I. Mikhailova, Tatyana A. II. Title. SB608.C7R6813 1992 634.9'75 – dc20 92-21023 CIP

Typesetting: Best-set Typesetter Ltd., Hong Kong
31/3145-5 4 3 2 1 0 – Printed on acid-free paper

Preface

Preservation of nature and the environment has become one of the most important issues of the end of the twentieth century. It has become evident that the methods used for industrial and agricultural production in many countries produce pollutants that cannot undergo natural neutralization by entering the atmosphere, soil or water. Ecosystems that have been developing for centuries are undergoing degradation and what is even more regrettable is that there is an actual threat of profound disorder in the biosphere which could lead to heavy and irreversible changes.

Fluorine derivatives are the most aggressive among toxic compounds polluting the atmosphere. Moreover, the percentage of fluorides in industrial emissions is constantly increasing with the bulk of fluorides being emitted by aluminium smelters. Fluorine is poorly detoxified by both plants and animals and the accumulation of even relatively low concentrations over a long period causes a cumulative toxic effect. Among woody plants conifers are less resistant to fluorine. Fluorine derivatives as phytopollutants have been studied less than sulphur compounds, nitrogen oxides, chlorine and hydrogen chloride. It was not until the late 1960s when there was a rapid decline of coniferous forests that researchers directed their attention towards phytotoxic properties of fluorides.

This book is the result of many years' study on the impact of fluorine on conifers. The work has been performed in Eastern Siberia where rapid development of the aluminium industry, which has arisen from the availability of electric power provided by hydroelectric plants, has rapidly become detrimental to coniferous forests in polluted areas.

The investigations have been carried out in the Laboratory of Pathology of Woody Plants in the Siberian Institute of Plant Physiology and Biochemistry, Siberian Branch of the USSR Academy of Sciences (Irkutsk, USSR). The laboratory first initiated the study as a single project. In 1973 the work was expanded and included in the projects of the Institute; some of the investigations have been further supported by the State Committee of Science and Engineering of the Council of Ministers of the USSR and a contract was signed with the Ministry of Forestry of the Russian Federation. The experiments were carried out at the laboratory field station in the Tunka valley (southern Baikal area) between 1976 and 1987. The station has experimental plots, permanent and transportable gas chambers with controlled levels of fluorine concentration and laboratory equipment suitable for conducting biochemical and toxicological assays.

The present work is one of the first to report a multipurpose study of the impact on woody plants of a single pollutant from industrial emissions; therefore, it is necessary to include in it some general problems encountered. The authors have tried, on the one hand, to write a complete book devoted to a narrow topic and, on the other hand (for practical purposes), to make each chapter and section independent and relatively complete. As a result, some problems will inevitably be discussed not once but several times from different points of view; however, such repeated discussions are not numerous.

The English edition of the book is a little different from the Russian publication in 1989. The authors have inserted some changes, made additions to Chapters 3 and 4 and eliminated Chapter 8. The basic data of Chapter 8 are discussed in Chapter 2.

Irkutsk, October 1992

A.S. Rozhkov
T.A. Mikhailova

Contents

Introduction: The Problem; Aims and Methods of Research

It has not been easy to accept the evident fact that our environment is more sensitive to industrial impact than are human beings and that this environment, with which man is constantly interacting, is a necessary requesite for existence. We have still not become aware of the importance of studying the impact of industrial pollution on ecosystems and the necessity to assess maximum permissible concentrations in the atmosphere of agents that are harmful to plants and plant communities. We also understimate the ability of protection zones around industrial plants to protect against local atmospheric pollution. Such zones are aimed not only at assuring safety to people but also at reducing the adverse impact of pollutants on local vegetation and animals.

Until the middle of the last century there were only occasional reports on the impact of smoke and gases on animal and plant organisms. The extent of environmental pollution was not large and was confined to smoke fires, waste gases from coal combustion and from small industrial works (foundries, soap works, soda factories etc.). Rapid increase in industrial emissions following the industrialization of Europe made environmental pollution become a serious ecological problem. From the middle of the last century the phytotoxicity and deleterious impact of pollutants on nature have begun to be more or less systematically investigated. Gas resistance of woody plants has been studied in Tarandt (Germany) where the Research Institute of Forestry was established. A monograph *Acid rains* by Smith was published in 1872 and in 1878 Hetz (cited by Hetz et al. 1985), a German scientist, described in detail visual symptoms of damage to green plants affected by sulphur dioxide (Hetz et al. 1985).

Evidence for the phytotoxic effect of industrial emissions has been rapidly accumulated since the beginning of the twentieth century. In Russia, the first investigations in this field were performed at the beginning of the century. Among them the most noteworthy are papers by Sabashnikov (1911a,b), in which the author considered at length the impact of coal smoke and tar fumes on woody plants. Systematic investigations of the resistance of plants to smoke began in the USSR in 1934 in the Laboratory of Plant Physiology and Biochemistry in the Department of Landscaping in the K.D. Panfilov Academy of Public Services (APS). In 1938 these investigations were continued at the Gorky State University. Interrupted by the Great Patriotic War the studies were resumed in 1946 in the APS by Krasinsky and co-workers. Krasinsky's team used the achievements of plant physiologists and the ideas of Timiryazev on photochemical reactions in plant cells to propose the theory of gas resistance of plants; this gave impetus to expanding research in this field.

At present, along with accumulation of descriptive data, great progress is being made in the development of experimental work and of comprehensive studies combining ecological, physiologo-biochemical and morpho-anatomical approaches to elucidate the problem of gas resistance of plants. Initially it was the toxic effect of sulphur dioxide on plants that was mainly investigated but since the 1970s there has been an increase in information on the phytotoxicity of other pollutants.

Industrial emissions can be classified as organic (various hydrocarbons, phenols, carbon disulphide etc.) and inorganic (acid gases, acid vapours, alkaline gases, oxides of metals etc.) substances. Acid gases and vapours are the most aggressive pollutants in terms of their toxic effects on plants. They are widespread and in most cases constitute the main part of emissions; for instance, world industry emits as much as 150 million tons of sulphur dioxide annually (Styrikovich and Vnukov 1981) but annual emissions of volatile organic compounds amount to 1 million tons (Kovalev and Polevaya 1981). The emission of sulphur dioxide by industrial works is predicted to be five times larger by the end of the century (Yager 1975) if no engineering progress is made.

In the 1960s and early 1970s the problem of reducing the adverse effect of enhanced concentrations of atmospheric fluorine on coniferous forests became most urgent. Central Europe and the USA were initial epicentres of the toxic impact of fluorine and therefore research efforts began here (McCune et al. 1964; Jacobson et al. 1966; Garber 1967; Robak 1969; McCune and Weinstein 1971; Dässler et al. 1972; Horntvedt and Robak 1975; Keller 1975a,b; Guderian 1979); fluorine was a major pollutant.

Atmospheric emissions containing fluorides are associated with a great number of industries. Industry has long employed inorganic compounds of fluorine: fluorspar, cryolite, fluorides of metals, hydrofluoric acid and boron trifluoride. The use of other fluorine-containing substances was limited by the availability of elemental fluorine. Much progress was made in the development of fluorine chemistry during World War II when the extraordinary properties of fluorine were used to solve problems of mastering nuclear energy; volatile fluorides of uranium allow uranium isotopes to be separated (Knunyants and Fokin 1963). In the last 30–40 years the rapid development of the chemistry of fluorine and its compounds has promoted fluorine from a little-known element to one that forms the basis of several large-scale industrial productions.

Chemists consider fluorine to be the most active and aggressive chemical element in nature. Its oxidative potential is higher than that of ozone. Fluorine does not exist in a free state in nature as it is very unstable and reacts immediately with air moisture to form hydrogen fluoride. Very little detoxification of fluorides can occur in plants, especially in conifers, which decline mainly because of the cumulative action of fluorides.

The process of stand degradation has expanded rapidly in recent years, with increasing involvement of fluorine. For example, in 1968 the forest area exposed to industrial emissions amounted to 400 000 ha in Europe, of which 300 000 ha were affected in Austria, 6000 ha in France, 15 000 ha in Rumania and 4000 ha in Sweden and Norway (Mathy 1977), but by 1984 the area of damaged forests amounted to 3.7 million ha in Germany alone (Bach 1985). In Austria, as reported at the 10th Austrian Conference on Forestry (Amschl 1986), 960 000 ha of forests were affected by atmospheric industrial pollution. By 1967 Poland lost about 240 000 ha of coniferous forests because of atmospheric pollution (Mowszowicz 1980), and by 1983 visual symptoms of the impact of emissions were observed over an area of 564 000 ha and latent symptoms were observed over 4 million ha (Fabijanowski 1986). Forest decline is in progress in Czechoslovakia where 10% of forests have already been lost and by the year 2000 this process could affect 40–50% of stands (Landa 1986). Further decline of forest is occurring in Sweden (Bengtsson 1986), France (Bonneau 1986), Canada (James and Courtin 1985) and the USA (Smith 1985). In the USSR forest decline from the impact of industrial emissions has not yet become disastrous; however, this is not the result of perfect purification systems but because such vast areas are covered by forests. For example, in Siberia as much as 400 000 ha of stands have declined because of the emissions from aluminium smelters and more than 0.5 million ha have been weakened to a varying extent.

The area of our studies is characterized by the predominant coniferous (taiga) forests, which are intensively used in large-scale pulp and paper plants and are increasingly used in aluminium production. Therefore, protection of forests against toxic effects of industrial emissions is a matter of great importance.

There are still many things not known about the effect of pollutants on single plants, plant communities and ecosystems. From the very beginning our investigations were multipurpose, aiming to elucidate general mechanisms of the effect of toxic compounds on the plant organism and to study the specific effect of fluorine on conifer pathology and resistance. Our next objective was to determine the relative phytotoxicity of acid gases, to show the effect on pollutant mixtures of synergism, addition and domination, and to examine the conditions controlling distribution of emissions from their source to the formation and structure of the polluted area.

Moreover, some special problems were examined including: phytotoxic activity of fluorine compounds; pathways by which pollutants enter the plant; translocation, localization and seasonal synamics of accumulation of pollutants in tissues; the influence on physiologo-biochemical processes;

specific features of the burn and toxic state; weakening and drying of the crown and of the entire tree; resistance of coniferous species to fluorine; and the role of dendrophilous insects in the decline of stands weakened by fluorine. Original and useful data have been obtained concerning most of these problems.

The basic practical task of the work was to assess maximum permissible concentrations (MPCs) of fluorine for conifers. To achieve this it was necessary to evaluate the general principles and methods of assessment of MPCs for plants as well as the use of additional parameters and correction factors. Also, a search for, and selection of, fluorine-resistant conifers was undertaken and consideration was made of their being used in protection zones around industrial works and settlements within the polluted area.

In this study natural stands have been examined in polluted areas and samples of air, water, soil and plants have been simultaneously analysed. The experimental part of the work has been performed at the field station of the Laboratory of Pathology of Woody Plants in the Tunka valley in the southern Baikal region. Here the experiments were carried out on naturally growing coniferous stands and in field chambers (permanent and transportable) that have controlled levels of pollution.

The construction of the basic chamber follows the design of a plastic greenhouse with a gabled roof. It consists of a framework covered with a polythene film and a base in the form of a frame set partially into the soil (Fig. 1a–c). The framework is hinged to the base so that it could be raised for total aeration. There are aeration holes in the front of the chamber, which also facilitate the insertion of polyvinyl chloride tubes for air sampling.

Four types of chambers were used; these differed in size, construction and function. (1) Transportable chambers of 2.5–3.0 m^3 (see Fig. 1b,c) were used mainly for performing physiological experiments. (2) Large permanent chambers of 20–25 m^3 (see Fig. 1a) were used to test tree resistance to fluorine. (3) Chambers with a wire (non-oxidizing) framework placed over crowns, and (4) Chambers mounted on tree trunks (Fig. 2a,b) were used to follow the pathways of fluorine into the plant, and fluorine translocation in tissues of trunk and branches.

In autumn, 10- to 15-year-old trees were transplanted into the base and were fumigated during the next growing season. Fumigation with hydrogen fluoride lasted for 18–20 h/day although in some experiments the duration was different. The required concentration of fluorine in chambers was produced by the flow of an air current over a vessel filled with hydrofluoric acid and supplied with an opening of a definite diameter. Large chambers were aerated with clean air supplied by a high-power fan. The maximum temperature variation inside and outside of the chambers was 11°C but, as a rule, it varied by 2 to 4°C; the maximum difference in air moisture reached 10% and averaged 4–6%. In the period of most intense solar radiation the chambers were shaded with white linen screens and during the hottest part of the day the chambers were opened and fumigation ceased. Samples of air

Fig. 1. **a** Permanent and **b**, **c** transportable gas chambers

from inside the chambers were taken daily with an aspirator to control the concentration of fluorine. Twice-distilled water served as an absorbent of soluble fluorides and an AFA-V-10 filter (USSR) was used to collect aerosol emissions (when samples were taken in the vicinity of the source of emission). The content of fluorine ions was determined photometrically and separately for soluble and solid fluorides.

Fluorine content was analysed in soil and plant samples (needles, phloem, sapwood and roots) following pretreatment, which involved ashing

Fig. 2. Local chambers **a** attached to the top of a 20- to 25-year-old tree and **b** a contact chamber on the trunk

(and melting for soil) and distillation with water vapours. The concentration of fluorine ions in the distillate was determined photometrically with the indicator xylenol orange (Kiseleva 1966; USSR Ministry of Public Health 1979).

The damage of needles was estimated visually by the extent of necrosis; 10% necrosis was considered to be a low extent of damage, up to 40% was intermediate and above 40% signified severe damage (Kulagin, 1974). Standard methods were used for statistical treatments of the material (Plokhinsky 1970).

Chapter 1

Fluorine as a Pollutant, Its Compounds in Industrial Emissions. Formation and Structure of Damaged Areas in Coniferous Forests

1.1 Acid Pollutants from Industrial Emissions and Role of Fluorides as Acid Pollutants

The effect of fluorine-containing emissions on vegetation have much in common with other acid pollutants. However, there are some specific aspects attributed to the ready distribution of this pollutant in the atmosphere, its high phytotoxicity and its prevailing cumulative effect on woody plants.

Ferrous and non-ferrous metal industries, oil refineries, pulp and paper plants, chemical industries, power plants consuming coal, and motor transport are referred to as the main sources of emission of acid gases into the atmosphere. Sulphur dioxide, hydrogen sulphide, nitrogen oxides, carbon monoxide, chlorine and hydrogen fluoride are the most severe pollutants for animal and plant organisms, although carbon monoxide is essentially non-toxic for plants.

The most widespread pollutants are sulphur dioxide and carbon monoxide because they are present in the emissions of most plants. Hydrogen sulphide emitted into the atmosphere is also transformed into sulphur dioxide by reacting with oxygen. More recently, nitrogen oxides have become more important as pollutants; this is not only because of the increase in industrial and motor transport emissions but also because of a relatively new source of nitrogen oxides, those from supersonic aviation. Halogens and their derivatives are less commonly met; however, their lower abundance is offset by their enhanced chemical aggression.

People and animals affected by acid gases have diseases mainly of the upper respiratory organs, may develop bronchial asthma and, in serious cases, have lung diseases. The effect of acid gases on plants results in the acidification of the cytoplasm and an ion imbalance, which affects reduction-oxidation reactions, inhibits photosynthesis and affects respiration and water metabolism. It initiates breakdown of polymer compounds and cell destruction. Irreversible changes in cells and organelles induce development of leaf necroses, reduction of assimilating area and malfunction of cambium. In woody plants all these events lead to suppression of growth and development, premature ageing and reduction of life. Prolonged exposure to acid gases is responsible for accelerated degradation of the plant organism and for decline in its resistance to deleterious biotic and abiotic factors, which leads to death.

Evidence from numerous observations and experiments suggest that fluorine and its volatile compounds and water-soluble salts have the highest phytotoxicity among acid pollutants (Thomas 1962; Ilkun 1971a; McCune and Weinstein 1971; Keller 1975a; Brennan and Rhoads 1976; Guderian 1979). Hydrogen fluoride (HF) is the most commonly met fluoride; it is emitted into the atmosphere mainly during production of aluminium, superphosphate, ceramics, glass, fluoroplastic and freon, as well as during decomposition of other fluoride compounds, including F_2, SiF_4 and H_2SiF_6.

At room temperature hydrogen fluoride, free of water, is a gas or liquid with a boiling point of 19.5°C; it produces fumes under the influence of air moisture. The binding energy of HF is much higher than in other halogens. Strong hydrogen bonds are formed between two HF molecules; therefore, in spite of the low degree of dissociation, an aqueous solution of hydrogen fluoride is a strong acid.

It is known that the higher the density of a gas and the more readily it dissolves in water, the higher is its toxicity to living organisms. In the case of hydrogen fluoride, its density is 20-fold over the density of hydrogen under similar conditions; it has a high affinity for water and is a powerful dehydrating agent (Saimons 1953). This accounts for the serious effect of hydrogen fluoride on plant and animal organisms. In addition, volatile fluorine compounds can aggregate in the air in considerable amounts and due to their high density in a gaseous form they tend to "stagnate" (Ishikawa and Kobayashi 1982).

We have attempted to estimate and compare the toxicity of the main acid gases for coniferous plants (Table 1). The toxicity of hydrogen fluoride was taken as 100 points (without regard for the high cumulative effect typical of fluorides) and a threshold concentration of fluorides was taken as the maximum content at which there was no visible sign of damage. The table is compiled from our own results and from a literature analysis (Tomas

Table 1. Toxicity of most commonly met acid gases for coniferous plants

Gas	Toxicity, points	Concentration of pollutants in needles (% DW)	
		Threshold	During decline
HF	100	0.004–0.006	0.020
Cl_2	0.3–30	0.10	0.25–0.40
SO_2	0.1–10	0.11–0.14	0.3–0.5
NO_2	0.03–3	–	–

1962; Ilkun and Bukolova 1974; Keller 1975a; Van Haut 1975; Ilkun 1978; Guderian 1979; Kulagin 1980; Hager and Kazda 1985; Smith 1985).

Relatively inert fluorine-containing compounds entering into the composition of solid aerosols are not harmless either. The toxicity of dust from aluminium smelters was estimated in experiments with common pine (*Pinus sylvestris* L.), yellow pine (*P. cooperi*), common larch (*Larix decidua* L.) spruce (*Picea abies* L.) and black alder (*Alnus glutinosa* (L.) Gaertn). Its effect is expressed as a reduction in linear growth and even in seedling death (Greszta et al. 1981). Metabolic disorders in plants exposed to cryolite have been described by Keller (1973).

1.2 Fluorides in Emissions from Aluminium Smelters and Their Accumulation in Polluted Areas

Of all branches of industry, the aluminium industry produces the greatest amount of fluorine-containing emissions. An electrolytic method of aluminium production from alumina when melted cryolite [Na_3AlF_6 (melting point, 980°C)] is used as an electrolyte involves the emission of a number of fluorides into the air. The ratio of gaseous and solid fluorides emitted is approximately equal. Gaseous fluorides are represented by hydrogen fluoride (HF), carbon tetrafluoride (CF_4) and silicon tetrafluoride (SiF_4), and solid fluorides by sodium fluoride (NaF), aluminium fluoride (AlF_3), calcium fluoride (CaF_2) and cryolite. Hydrogen fluoride is the most abundant emission of all these. According to Ennan and Takhtarova (1978), the leakage of fluorine is 20–30 kg per ton of smelted aluminium.

Current technological systems for the purification of aluminium smelter emissions can collect up to 98% of the fluorine discharged in gases and aerosols; however, an actual coefficient of purification is 30–75%; this value does not include waste through ventilation systems, which sometimes contain more fluorine than waste discharged through chimneys and aspiration systems.

Soil constantly acts as an absorbent of pollutants. The inherent level of fluorine in soils from different localities varies from 0.003 to 0.032% (Vinogradov 1957). An enhanced level of fluorine may affect the plant both by deterioration of properties of the soil itself, expressed as a change in structure, water permeability and biological activity, and by adverse effects on metabolism after entry into the plant.

In soils exposed long-term to industrial emissions, the upper soil horizon undergoes the greatest changes. It is the place where the majority of precipitating solid particles, dissolved salts and absorbed gases are localized. Nevertheless, some compounds, e.g. those containing sulphur, may migrate as deep as 1 m or more (Krouse et al. 1984). Our data and that of other authors (Ilkun and Motruk 1976; Löffler 1983) demonstrates that fluorine compounds accumulate predominantly in the upper 5–10 cm of the soil. The amount rapidly decreases with depth and in a layer at 40–50 cm background levels are usually found.

We have analysed soil samples from the upper horizon of soil profiles. The results revealed a distinct relationship between the fluorine accumulation in soil and the distance from the source of emission (Fig. 3). The highest content of fluorine was found in the zone of highest concentration of emission; it exceeded by 18- to 20-fold the content of the control. With a greater distance from the source of emission the amount of fluorides in soil decreased but was still 1.5- to 1.8-fold higher than the control, even in the zone of low contentrations. A high level of fluorine in the roots of trees near to the smelter seems to result from the impact of not only atmospheric but also soil fluorides.

The capacity to retain fluorine depends on the soil type. Ilkun and Motruk (1976) inferred that it is least in river sand. With increasing slime and colloid fractions in soil, the fluorine-retaining capacity increases and is highest in grey forest soil and in chernozem. The same authors found that saturation of soil with fluorine was accompanied by soil destruction and a decrease in water permeability. Leaching of this element from soil is followed by simultaneous removal of potassium and sodium, while calcium remains almost immobile. Under normal conditions only a small part of soil fluorides are available for plants as fluorine is incorporated into the aluminosilicate complex in acid soils and reacts with calcium and silicon to form insoluble salts in alkaline soils (Tarchevsky 1964).

Snow cover plays an important role in fluorine uptake during winter. Analysis of snowfall for a month in the zone of high concentrations of atmospheric fluorides showed that the content of soluble fluorine compounds in the snow (on conversion to fluoride ions) amounted on average to 4.7 mg/litre. This concentration is almost four times greater than the

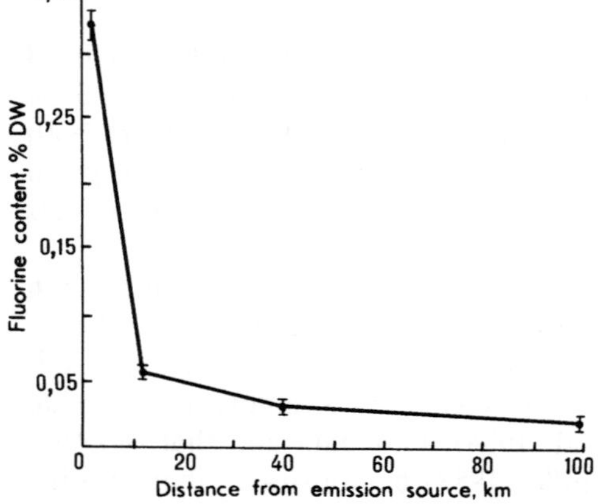

Fig. 3. Fluorine content in the upper soil layer at different distances from the aluminium smelter

inherent fluorine content in the hydrosphere (1.3 mg/litre) and is three times greater than the maximum permissible concentration of this element (1.5 mg/litre) in drinking water in the USSR.

1.3 The Effect of Mixing Hydrogen Fluoride with Other Acid Pollutants

The ambient atmosphere around many industrial plants, including aluminium smelters, is characterized by the presence of a complex of pollutants acting simultaneously. Under such circumstances one can expect variation in the phytotoxic activity of emissions. The effect of a mixture of pollutants on plants is a relatively new field of research and not much data are yet available to characterize the effect of combined pollutants. However, such research is needed to assess MPCs of air pollutants and to advise businesses within a certain region of allowable levels of emission. This would reduce the damage suffered by vegetation, especially sensitive coniferous forests. There are many examples where factories emitting pollutants of the same type in one place caused massive decline of the vegetation and resulted in the formation of vast industrial deserts.

Four types of action of pollutant mixtures can be distinguished; addition, synergism, domination and antagonism. Addition describes the situation where the extent of damage caused by a pollutant mixture corresponds to the sum of damages caused by each substance. Synergism implies a greater overall damage than the sum of single damages, and antagonism implies less damage by substances acting together than by each substance acting separately. The domination of any one pollutant is the case when the extent of damage by a mixture containing this pollutant remains the same as under exposure to this pollutant alone.

The influence of any pollutant mixture on the plant organism largely depends on the concentration, duration and sequence of the mixture as well as on physico-geographical conditions. In addition, gas resistance of plants is important as is their age and stage of development. For example, the same mixture composed of low concentrations of sulphur dioxide, nitrogen dioxide and ozone had differing effects on eight clones of eastern white pine with different gas resistances. Even within a group of clones of similar sensitivity three types of effect were noted: additive, synergetic and antagonistic (Yang et al. 1982).

Analysis of available data suggests that synergism is more common when plants are exposed to high concentrations of pollutant mixtures. Addition and domination effects are possible with the low concentration of acid gases in the atmosphere. Combinations of pollutants resulting in decreased harmful activity are extremely rare. Antagonism might be expressed when acid and alkaline gases, or oxidants and reducing agents are simultaneously present in the air.

Antagonism of mixtures of sulphur dioxide with ozone and of nitrogen oxide with ozone has been reported by some authors (Landolt and Keller 1985). The addition of sulphur dioxide caused a reduction of the adverse effect of the mixture of ozone and products of petrol consumption (Haagen Smit et al. 1952). The decline in phytotoxicity of sulphur dioxide and phenol following the addition of a low concentration of pyridine has been reported (Tarabrin and Bashkatov 1986).

However, in the majority of cases the action of various combinations of similar pollutants results in development of additive and synergetic effects. When oats, barley and lucerne were fumigated with acid gases and mixtures of two pollutants no antagonism was recorded; most common was the additive effect (Bennett and Hill 1974). Pronounced synergism was reported in the cases of mixing sulphur dioxide with hydrogen chloride (Guderian 1979), oxides of nitrogen (Bennett and Hill 1974) and ozone (Menser and Heggestand 1966; Boyer et al. 1986).

Bearing in mind the strong aggression of gaseous fluorides, one can theoretically predict the synergetic effect of fluorine in combination with other pollutants. The prediction is corroborated by data showing a marked increase in damage suffered by agricultural crops on exposure to a mixture of hydrogen fluoride and sulphur dioxide (Ten Houten 1974; Morel and Chaouard 1967).

One of the aims of our work was to evaluate comparatively the toxicity of the most common acid gases and their mixtures upon coniferous plants. We also followed the effects of combinations of pollutants, i.e. the demonstration of synergetic or additive effects as well as those of antagonism and domination.

The experiments were run in field chambers with 10- to 12-year-old Siberian larch trees as the objects of the study. The following gases were used for fumigation: (1) hydrogen fluoride; (2) chlorine; (3) sulphur dioxide; (4) carbon monoxide; (5) nitrogen oxides (NO and NO_2); (6) hydrogen fluoride and chlorine; (7) hydrogen fluoride and sulphur dioxide; (8) hydrogen fluoride and carbon monoxide; (9) hydrogen fluoride and nitrogen oxides; (10) chlorine, sulphur dioxide, nitrogen oxides and carbon monoxide; and (11) hydrogen fluoride, chlorine, sulphur dioxide, nitrogen oxides and carbon monoxide.

The gases were obtained by reaction of weighed amounts with an excess amount of a certain acid. Hydrofluoric acid was a source of hydrogen fluoride. The concentrations of gases, except for hydrogen fluoride, were determined in chambers with a universal field gas analyser UG-2 (USSR). The content of hydrogen fluoride was assayed photometrically. The following concentrations of pollutants were used in the experiments: 0.1–0.2 mg/m^3 HF; 1.0 mg/m^3 SO_2; 1.0 mg/m^3 Cl_2; 2.0 mg/m^3 $NO_{(2)}$; and up to 30 to 35 mg/m^3 CO. The experiments were repeated five times. It was predetermined that a given concentration of a gas was maintained for 2–3 h in the chambers, which were 2 m^3 in volume, and than the concentration began to drop; therefore, reacting substances where regularly renewed in a reaction vessel.

The plants were fumigated for 18 h/day. In the warmest part of the day, from 12 a.m. to 6 p.m., fumigation was stopped. The fumigation continued for 20 days. During this period the larch needles treated with 0.1–0.2 mg/m^3 hydrogen fluoride suffered total damage. The damage to plants was estimated visually by assessing the number of necrotic needles in the tree crown (as a percentage of the total number of needles) (Table 2). Gases tested could be arranged in the following order in terms of declining toxicity to plants: HF, Cl_2, SO_2, $NO_{(2)}$ and CO.

No visible signs of damage were observed in larch fumigated with nitrogen oxides and carbon monoxide. It is known that carbon monoxide can be oxidized to carbon dioxide and then incorporated in the photosynthetic cycle. Nitrogen oxide is oxidized in air to become nitrogen dioxide or can be converted into gaseous nitrogen by means of photochemical reactions. Nitrogen oxide dissolves in water to form nitrite and nitrate ions which can be reduced to ammonia in leaf cells (Bennett and Hill, cited by Smith 1985). Ammonia interacts with ketonic acids to form amino acids (Kretovich 1980). The plant is also able to detoxify other acid gases. Thus, sulphur dioxide can be either oxidized to sulphate, which reduces its toxicity 30-fold (Tomas 1962), or reduced to hydrogen sulphide and incorporated into amino acids. We failed to obtain evidence for specific mechanisms of detoxification of chlorine and fluorine in plants. These elements are apparently neutralized by buffer systems in the cell (see Chapter 3).

When hydrogen fluoride is added with chlorine or sulphur dioxide plants suffer greater damage, even during a shorter period of fumigation. In other

Table 2. The degree of visual damage of needles of 10- to 12-year-old Siberian larch trees by acid gases and gas mixtures

Gas (mixture)	Concentration (mg/m^3)	Duration of fumigation (days)	Amount of necrotic needles (%)
HF	0.1–0.2	18–20	90–100
Cl_2	1.0	20	Up to 60
SO_2	1.0	20	Up to 40
$NO + NO_2$	2.0	20	No damage
CO	Up to 30–35	20	No damage
$HF + SO_2$	0.1–0.2 + 1.0	3	80–100
$HF + Cl_2$	0.1–0.2 + 1.0	2	80–100
$HF + NO_{(2)}$	0.1–0.2 + 2.0	20	90–100
HF + CO	0.1–0.2 + 30.0	20	90–100
$Cl_2 + SO_2 + NO_{(2)} + CO$	1.0 + 1.0 + 2.0 + 30.0	7–8	80–100
$HF + Cl_2 + SO_2 + NO_{(2)} + CO$	0.1–0.2 + 1.0 + 1.0 + 2.0 + 30.0	0.5	80–100

words, pronounced synergism is observed with exposure to mixtures of hydrogen fluoride and chlorine, or hydrogen fluoride and sulphur dioxide. When the plants were fumigated with mixtures of hydrogen fluoride and nitrogen oxides, or hydrogen fluoride and carbon monoxide there was no increase in damage and the extent of damage remained the same as for exposure to hydrogen fluoride alone. Hence, in this case, hydrogen fluoride has a dominating toxic effect. The fumigation of plants with a mixture of chlorine, sulphur dioxide, nitrogen oxides and carbon monoxide showed this mixture to be several times more toxic than chlorine and sulphur dioxide acting separately. A mixture of all the gases (hydrogen fluoride, chlorine, sulphur dioxide, nitrogen oxides and carbon monoxide) turned out to be the most destructive to plants. Needles exposed to this mixture die during 10–15 h. In this case several synergetic effects, produced by mixtures of hydrogen fluoride and chlorine, hydrogen fluoride and sulphur dioxide, and chlorine and sulphur dioxide, become superimposed.

Thus, the fluorine-containing emission is the most dangerous to plants in the presence of atmospheric chlorine and sulphur dioxide. Hydrogen fluoride is a dominant pollutant in the presence of nitrogen oxides or carbon monoxide.

During experimental fumigation of trees with acid gases and gas mixtures we also found certain differences in the development of visible damage to the needles. The effect of hydrogen fluoride and sulphur dioxide on larch initially induces development of chlorosis on apices of needles of auxiblasts. Chlorosis spreads through the needles, whose apices become brown or greyish-brown because of cell death. Then necrosis rapidly extends to all the needles of auxiblasts. A little later the needles of brachyblasts are damaged in the same manner and the damage spreads to all needles of the tree. Dead needles gradually fall from the tree. This process proceeds more rapidly in the presence of hydrogen fluoride and gas mixtures containing hydrogen fluoride. Fumigation with sulphur dioxide results in a longer period between the first appearance of chlorotic mottling and the death of needles; damaged needles are more lightly coloured than under exposure to hydrogen fluoride.

Other symptoms appear following the exposure of larch to chlorine. Needle colour also changes though there is no intense discoloration. First the needles of auxiblasts and then those of brachyblasts become grey-green and later on light grey. Owing to extensive dehydration the needles dry out and break readily. They gradually turn brown and finally become greyish brown. Initial symptoms of the damage caused by mixture of chlorine with sulphur dioxide, nitrogen oxides and carbon monoxide are the same as for chlorine alone, though in the last stage of dieback the needles turn dark brown.

The elucidation of specific symptoms of plant damage by pollutants is of considerable interest to researchers as it can be useful for diagnostic purposes. Our experiments revealed that certain specific visual symptoms of plant damage by acid gases only become apparent with exposure to high concentration of the pollutant. It is harder to characterize symptoms that result from long-term exposure to relatively low concentrations of acid gases, where it is difficult to pinpoint an injurious factor.

1.4 The Spread of Fluorine from Emission Source

The character of toxic compounds and the distance they spread from the source of emission is determined by a number of factors, the main ones being the amount and state (gaseous, aerosol) of a pollutant, the height of chimneys, the presence of additional waste, the frequency and velocity of winds from different directions, and the relief and extent of afforestation of the surrounding area.

As already stated, aluminium smelters emit into the atmosphere approximately equal amounts of gaseous (HF, CF_4, SiF_4) and solid (NaF, AlF_3, CaF_2, Na_3AlF_6) fluorides. Gaseous fluorides are more toxic though the bulk of them is transported over shorter distances than solid fluorides. The concentration of gaseous fluorides maximally exceeding that of aerosols is within 4–12 km from large-scale smelters. As distance increases the amount of gaseous fluorides declines and by a distance of 30 km they only amount to 10–15% of the total content of fluorine in the polluted air current (Fig. 4).

Forests play an extremely important role in limiting distribution of industrial emissions. The mechanisms of this limitation are intricate. In addition to a filtering function, forests affect the velocity and direction of wind; this is one of the major factors responsible for formation of the polluted zone.

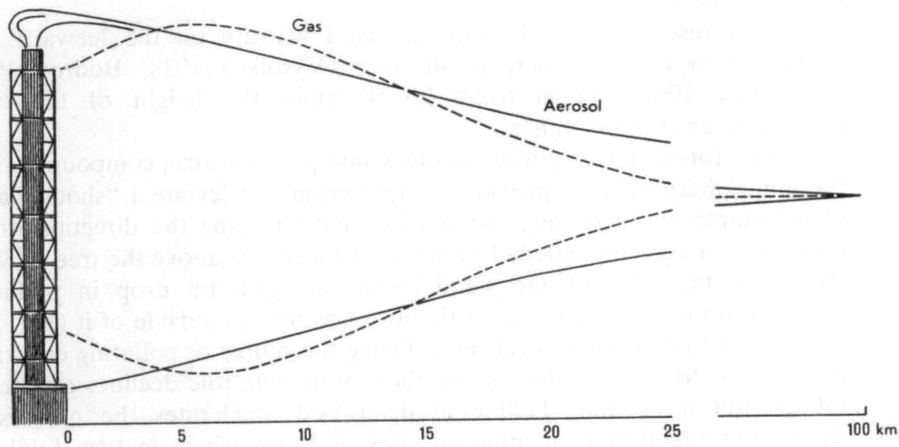

Fig. 4. The ratio of gaseous and solid fluoride (aerosol) emissions

Table 3. Wind velocity in and above the forest

Place of measurement	Height above soil (m)	Average velocity (m/s)
Above crowns	16.9	1.61
Upper boundary of crowns	13.7	0.90
Inside of crown canopy	10.6	0.69
Underneath crowns	7.4	0.67
In the middle between canopy and soil	4.3	0.69
Over soil	1.1	0.60
On soil surface	0.0	0.0

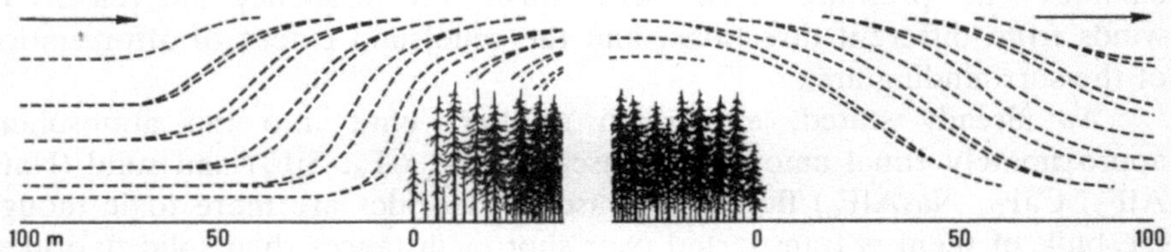

Fig. 5. Direction of wind flow at the exposed forest edge and over an open area on the leeward side

According to Nesterov (cited by Molchanov 1961) the velocity of wind decelerates by 20–50% as far as 60 m from the forest edge, depending on thickness and height of the stand, the presence of undergrowth and the actual wind velocity. At a certain distance within the stand, the wind stops blowing and changes to a barely perceptible air current or complete calm. Giger (cited by Tkachenko 1952) reported data on wind velocity in a pine forest (Table 3).

The forest causes a decrease in wind velocity on the leeward side extending over a considerable distance (Vysotsky 1938; Bodrov 1940; Molchanov 1961), which totals 20–50 times the height of the trees depending on stand thickness.

Thus, forest not only absorbs dusty and gaseous toxic compounds from the atmosphere but also provides a mechanism to alleviate a "shock wave" of pollutants by decreasing the velocity and changing the direction of air currents. Air currents repelled by a wall of forest rise above the tree crowns; therefore, the edge of the stand causes a significant drop in pollution concentrations within the rest of the forest as well as outside of it (Fig. 5).

Trees situated at the forest edge facing the source of polluting emissions are the first to die. Because of this their protecting role declines drastically though not altogether. Felling of deadwood accelerates the process of weakening and drying of adjacent sites of forest which in turn fulfil the windbreak function. In coniferous forests, especially larch ones, the first layer of trees to die protect other members of the species under the same canopy for many years.

For about 20 years the question has been discussed in the literature of whether drying and dead standing trees should be felled at forest edges in zones of heavy air pollution. German foresters Liebold and Ranft (1981) state that until now there was a tendency in spruce-forest management to create an impression of "clean, healthy forests" by removing all dead or drying trees but, at present, they tend to use the potential protecting properties of this part of the forest to prevent the damage of the rest of the forest. It has been shown that any unnecessary felling or removal of dried trees from forest edges facilitates the spread of damage into the forest's inner part which was still relatively unaffected by pollutants.

1.5 Formation of Zones that Differ in Fluorine Concentration and Extent of Conifer Damage

The area of pollution around the source of emission can be divided into three zones based upon; (1) the concentration of toxic compounds in the air, and (2) the degree to which plants are exposed to the toxic compounds. The dimensions and length of formation of such zones differ greatly in these two cases, especially for pollutants that can be accumulated in plants in lethal doses even under conditions of moderate concentration within the environment.

First, we will examine the distribution of different concentrations of fluorine over an area around aluminium smelters. Most authors distinguish zones of high, moderate and low concentrations of pollutants on maps of distribution of emissions. They are also referred to as 1st, 2nd and 3rd polluted zones respectively; we have used this system in our previous papers. Such a classification of polluted areas is fairly arbitrary but is nevertheless useful, especially in the early years of pollution. Maps with a large number of isobars of pollutant concentrations can be compiled using long-term samplings of air from various sites.

Ideally (in a flat country uniformly covered with forests of the same species and exposed to similar winds from all directions and of similar velocity) zones and isobars indicating pollutant concentrations would appear as concentric circles or belts. In reality all three zones have more or less amoeba-like configurations whose shapes depend on physico-geographical features of the country, such as landscape, meteorological conditions and type of afforestation. In mountain-forest steppe and mountain-taiga regions of Siberia the zones of different fluorine concentrations around aluminium plants are of a mosaic pattern. The zone of high concentration has, in addition to a central part, some detached spots which represent hill slopes facing the smelter; these spots are within the zone of moderate concentration which can, in turn, extend as narrow arms or single spots into the zone of low fluorine concentration.

Up to now no attempts have been made to quantify atmospheric fluorine levels according to zone. In our studies we used the following concentration ranges of gaseous fluorine: 1st zone (0.6–0.05 mg/m^3); 2nd zone (0.04–0.012 mg/m^3); and 3rd zone (0.011–0.0006 mg/m^3). Solid compounds of fluorine were not considered as they are of relatively low toxicity to plants, although they were present in large concentrations even in the periphery of the polluted zone. Concentrations of fluorine found in the zone of young needle burn (the index of the highest concentration is taken from Kazantseva 1966) are considered to be high; moderate concentrations correspond to the zone where needles are damaged only on some trees and by not more than 20% during the first 4–5 years of smelter operation (up to the appearance of a cumulative effect at the level of the whole plant); and low concentrations failed to induce any visible injury to trees over the same period.

According to our evidence the following pattern of fluorine distribution is typical for an area around large-scale (20 buildings and more) aluminium plants situated in the south and middle taiga of Eastern Siberia: the 1st zone extends radially from the smelter in various directions up to 4–6 km, the 2nd zone up to 10–15 km, and the 3rd zone up to 40 km. In heavily broken mountainous country an enhanced content of gaseous fluorine (up to 0.005–0.008 mg/m^3) is sometimes recorded as far as 100 km from the smelter. In moderately broken mountainous country industrial emissions are spread over less distance: up to 3 km in the 1st zone, up to 7 km in the 2nd zone and up to 15 km in the 3rd zone (Surtaev 1985). The boundaries of all zones expand with forest decline.

The process of forest decline is nearly complete after the first decade in the zone of high fluorine concentration but it takes two to three decades in the zones of moderate and low concentration. The boundary line of forest decline becomes visible in 12–20 years. In this period one can distinguish in the polluted area zones corresponding to the three extends of coniferous tree damage and we consider the following terms to be appropriate descriptions: (1) the zone of lethal accumulation of fluorine by trees, (2) the zone of chronic disease of trees, and (3) the zone of insidious (latent) weakening of trees (Fig. 6). These zones, as already mentioned, do not correspond to the 1st, 2nd and 3rd zones. The zone of lethal accumulation of fluorine in trees embraces the zones of high, moderate and a part of the low concentrations of fluorine in air. The duration of its formation, its total area and extent of penetration into the zone of low concentration correlates with the amount of emission and the type of country.

The formation of the zone of lethal accumulation of fluorine is a prolonged process. It originates at a certain distance from the source of emission as discrete areas on the windward side of forest edges in uplands. The possibility of further expansion of the zone depends primarily on the

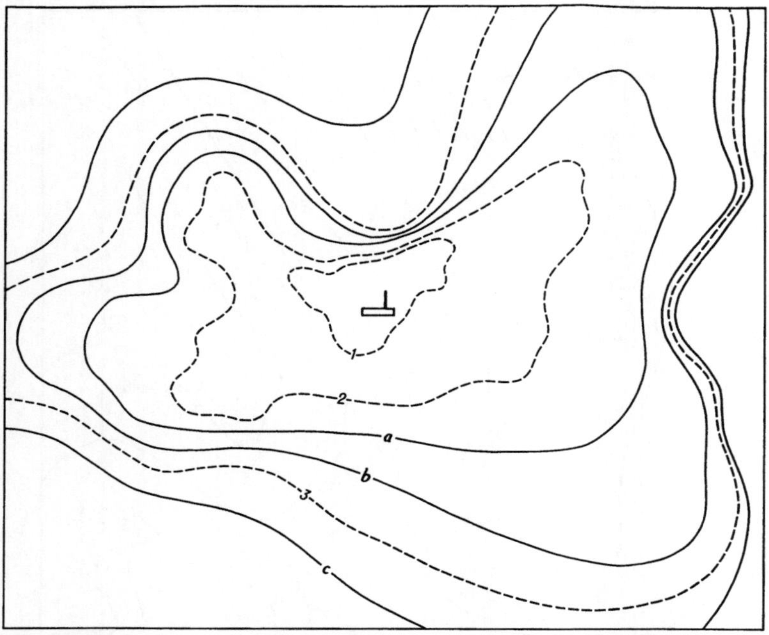

Fig. 6. A simplified scheme of the dimension of zones with different atmospheric fluorine concentrations (*1–3*) and different degrees of conifer damage (*a–c*) around the aluminium plant (see text)

relief of the land (see Fig. 8). The relief affects movement of local air currents and therefore also affects local tree damage following short-term exposure to high fluorine concentrations or fluorine accumulation in tissues and organs of trees. The role of this cumulative effect becomes more important with the distance from the source of emission, while differences in tree damage at different heights above sea level become less significant. This is particularly noticeable at the boundary line between the zones of chronic disease and latent damage of trees.

In the low-lying land of rivers and creeks, even in the close vicinity of the smelter, spruce, spruce–larch and pine stands can survive for 20–30 years. Mature and old trees decline during this period and young ones become slightly damaged according to external parameters, such as linear and radial increments of the trunk and needle length. This can be attributed to several factors, with landscape and meteorological conditions being of primary importance. These can provide favourable conditions for vertical diffusion of air which decreases the concentration of fluorine to a minimum value. An important role in retaining the resistance of trees and assisting with detoxification of the environment is also played by edaphic factors including soil moisture.

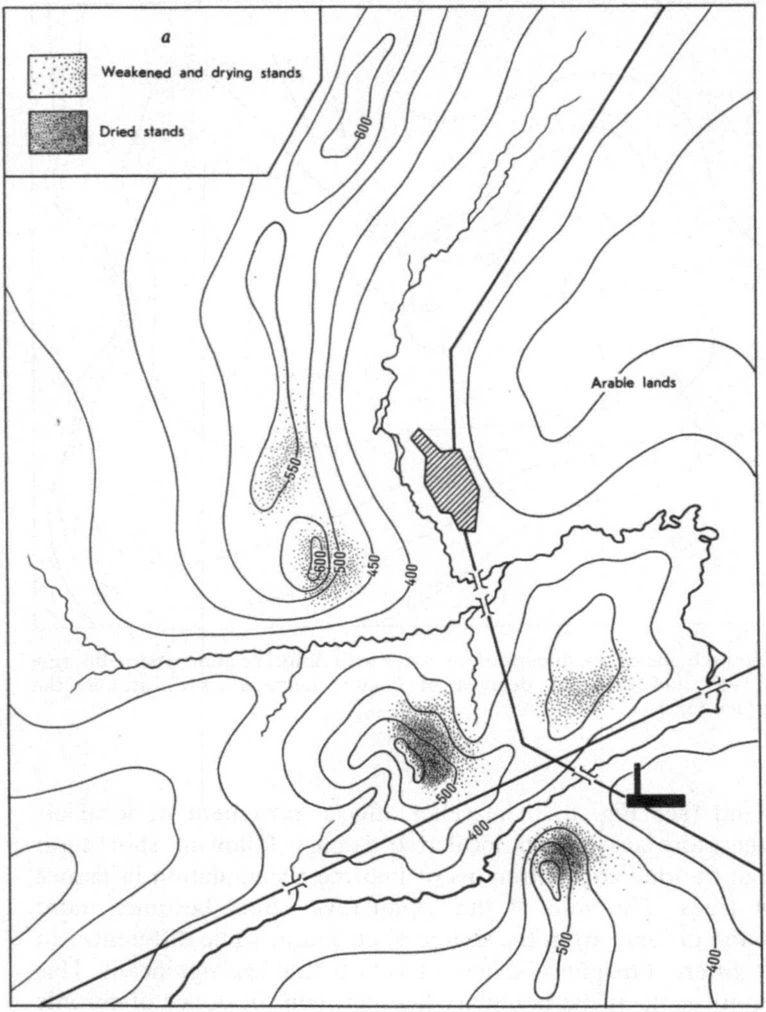

Fig. 7 a–d. A principal scheme of the formation of zones of lethal fluorine accumulation in conifers: **a**, **b**, **c**, **d**, show the condition of stands following 5, 10, 15 and 20 years respectively of operation of the aluminium smelter

Examination over many years of the decline in coniferous forest related to different zones around aluminium smelters allows the compilation of basic schemes on the formation of the zone of lethal accumulation of fluorine by trees (Fig. 7a–d). Kondratov, a forest pathologist, participated in the compilation of such schemes.

Within the whole zone of lethal accumulation of fluorine, especially in its periphery, degraded coniferous stands are gradually substituted by young

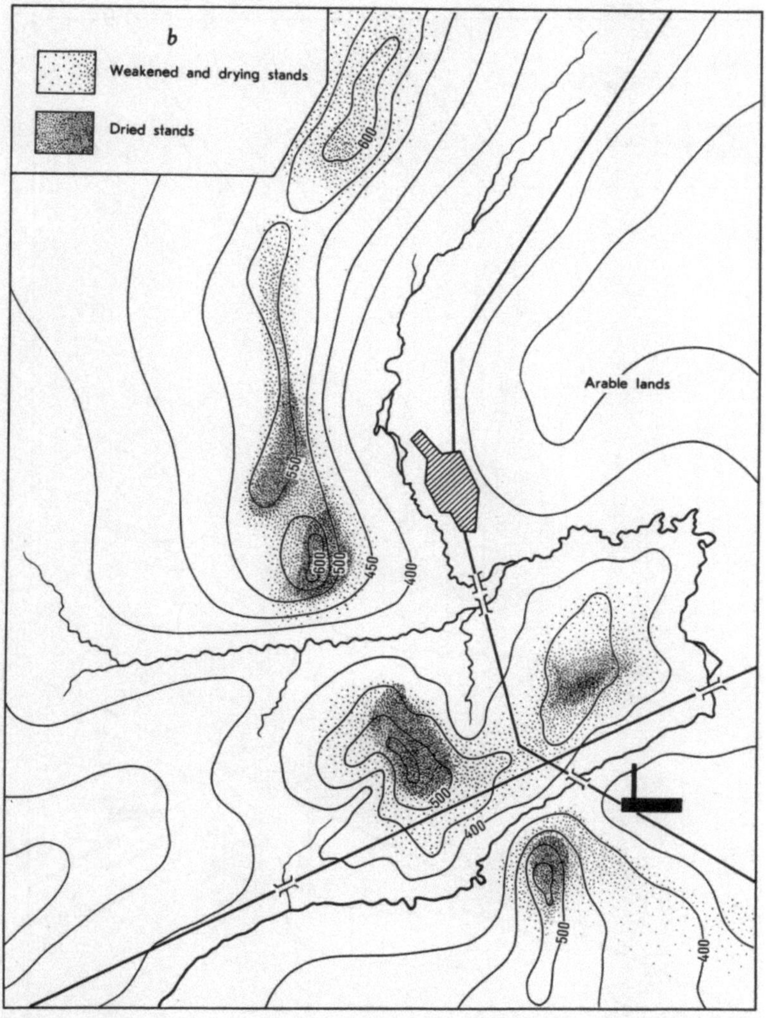

Fig. 7 (*Continued*)

deciduous forests consisting mainly of birch and aspen with reafforestation of conifers under their canopy. The performance of the young trees is normal until they reach the first layer of trees and thereafter, as early as 40–60 years of age, they begin to decline and fail to produce seeds.

The zones of chronic disease of trees do not conform to either "chronic damage" of plants by gases, which has been used in the literature since the end of the last century, or the term "chronic disease" of plants suggested

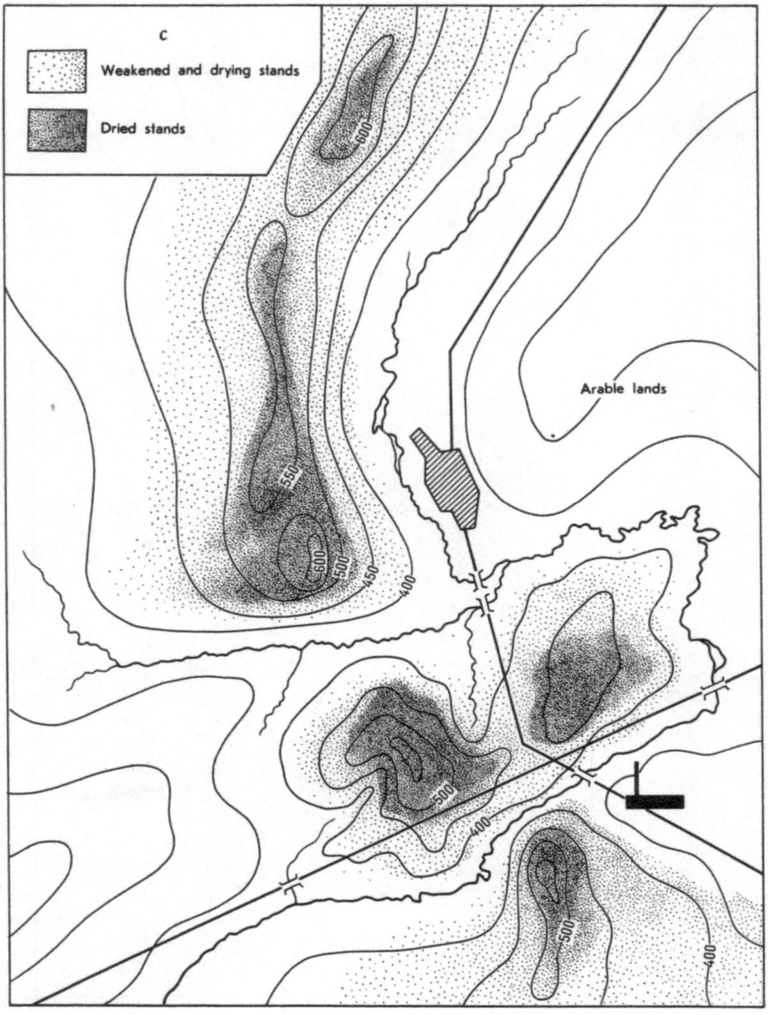

Fig. 7 (*Continued*)

by Wentzel (1967). These terms imply long-term accumulation in plants of the injurious factors which induce death and are distinct from the "acute damage" which results in rapid, sometimes instantaneous, death. Such descriptions of damage are inadequate for woody plants since they need a relatively long time to accumulate fluorine to a critical level and to decline, even under conditions of high concentrations of hydrogen fluoride. In the scheme proposed the zone of chronically weakened trees is characterized by reversibility of their damage with a general reduction in protection and

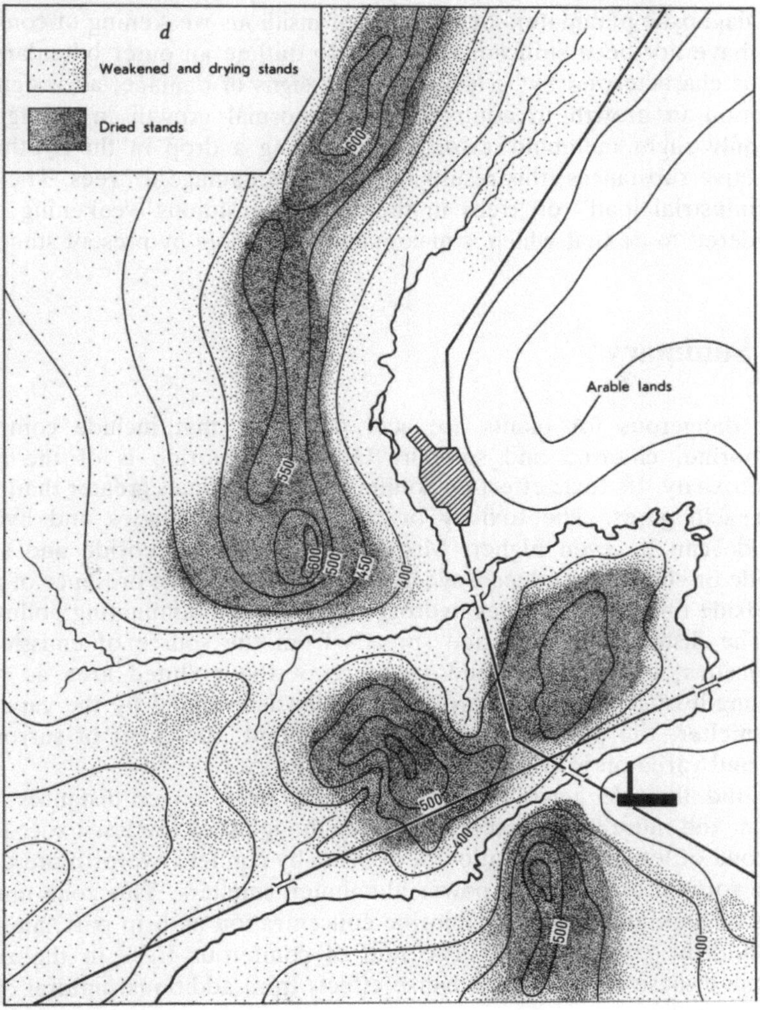

Fig. 7 (*Continued*)

repair abilities; the stand ecosystem is disturbed in its most sensitive steps (for example, changes in the composition of dendrobiont insects and fungi, and a reduction or disappearance of lichens) but remains basically steady. The fate of the forest depends here almost completely on the adverse effect of additional natural or man-made factors, such as an increase in emissions, disturbances in the level of ground water, outbreaks of pests or drought. Thus the zone of chronic disease should be considered as the zone of extreme risk.

Diagnostic parameters of the zone of insidious weakening of coniferous trees have not been sufficiently studied to outline an outer boundary. The zone is characterized by: a lack of visible signs of damage; an insignificant reduction in growth in adult trees and normal growth in young trees; and only slight metabolic disorders, including a drop in the synthesis of protective substances; this results in secondary damage to trees. The extent of "industrial load" on trees in the zone of insidious weakening can be considered to be that which is maximally permissible by present standards.

1.6 Summary

Most dangerous for plants are acid pollutants that include compounds of fluorine, chlorine and sulphur. Hydrogen fluoride is of the highest phytotoxicity. Its toxic effect on conifers is 3–1000 times greater than that of other acid gases. The toxicity of mixtures of acid gases and hydrogen fluoride can be even higher. Mixtures of hydrogen fluoride and sulphur dioxide or chlorine produce a synergetic effect and in the presence of carbon monoxide or nitrogen oxides hydrogen fluoride is a dominating pollutant.

The distance that fluoride spreads from the source of emission, the dimensions, configuration and structure of the polluted area as well as the length of its formation are all basically determined by the capacity of the smelter, the physical state of the pollutant, the relief of surrounding land and variation in wind direction and velocity.

Solid fluoride aerosols are transported over longer distances (up to 100 km and more) than gaseous fluorides though they are less toxic. It takes the zone of lethal accumulation of fluorine by conifers two or more decades to be formed around high-power aluminium smelters. This zone embraces areas of high and moderate fluorine concentration and, in part, an area of low fluorine concentration. The fate of coniferous trees in the zone of chronic disease depends on adverse effects from additional natural or man-made factors.

Chapter 2

Resistance of Conifers to Fluorine. Tree Decline

2.1 General, Species and Individual Resistances

The lower resistance of conifers to acid pollutants as compared to the resistance of deciduous trees is best illustrated by fluorine. Fluorine phytotoxicity is determined by the effect of burning and the toxic effects of fluorine accumulation in organs and tissues. The enhanced sensitivity of conifers to hydrogen fluoride is related to long needle life. Accumulation of hydrogen fluoride in the needles over many years results in extensive damage of the crown, reduction of the area of assimilation and finally plant death because of carbon starvation. Moreover, the high sensitivity of evergreen conifers arises from a relatively poor development of storage tissues and, consequently, insufficient accumulation of storage substances as well as a low regenerative capacity.

Under conditions of air pollution it is coniferous stands with a prevalence of pine, spruce, fir and Siberian cedar pine that suffer greater damage (Abramashvili 1957; Dässler et al. 1972; Keller 1975a; Brennan and Rhoads 1976). The reduction in needle life, in terms of weight loss, burn and decrease in length, a resultant thinning of the crown, and the decline in linear and radial growth are the most commonly observed traits of conifer injury. Exposure to smoke reduces formation of seed cones in conifers, affects the size of seed cones and seeds, and also interferes in seed germination (Popovichev 1980).

The data on relative gas sensitivity of conifers are rather contradictory. Antipov (1975, 1976, 1978) used the "expert's assessment" method to classify coniferous species according to their resistance to gases; the following ones are considered to be the most resistant: Sukachev's larch (*Larix sukaczewii* Del.), Japanese larch (*Larix gmelinii* (Rupr.) Rupr. var. japonica (Maxim. ex Regel) Pilger), Siberian larch (*Larix sibirica* L.), grey pseudotsuga (*Pseudotsuga caesia*), prickly spruce (*Picea pungens* Engelm.), Engelmann spruce (*Picea engelmannii* Engelm.) and Austrian pine (*Pinus nigra* Arnold.). The author stated that it was difficult to classify particular species reliably because of considerable variation in approach as well as divergent ecological conditions and tree state. It is more commonly recognized that all spruce species, among them the prickly spruce, belong to gas-sensitive species (Sadilova 1964; Greshta 1970; Dässler et al. 1972; Kluczyński 1976; Lenzburg 1985). Siberian fir (*Abies sibirica*) and other species of the same genus are considered to be highly sensitive to air pollution (Sadilova 1964; Antipov and Bolotov 1977) and it is recommended that this species should not be used to landscape industrial centres within a radius of 20 km of the industry.

The common larch (*Larix decidua* Mill.) is considered to be more sensitive to gases than spruce, pine and fir (Adams et al. 1957; Rohmeder and Schönborn 1965; Wentzel 1968; Dässler et al. 1972), while other authors (Kisser 1964, cited by Garber 1967; Horntvedt and Robak 1975) believe it is fairly resistant to gases. Examining gas-absorbing capacity of conifers, Getko and coworkers (1978) concluded that larch forests were more resistant to gases under constant low pollution and compared to pine forests they accumulated twice as much sulphurous compounds in their needles, the total area of which was three times greater. The evidence for gas resistance in pine is also contradictory. Some authors (Rohmeder and Schönborn 1965; Wentzel 1968; Horntvedt and Robak 1975; Kluczyński 1975) believe the pine to be moderately resistant, while the majority of workers consider it to be very sensitive to gases (Sadilova 1964; Swieboda 1964; Garber 1967; De Cormis 1970; Dässler et al. 1972; Horntvedt and Robak 1975; Jamrich 1976; Kluczyński 1976).

We propose our own assessment of relative fluorine resistance of coniferous trees in Siberia based on experiments carried out over many years in field gas chambers, examination of hydrogen fluoride polluted stands and biochemical assays. According to us, decline in conifer resistance may be ranked as follows: Dahurian larch (*Larix gmelinii* Rupr.), Siberian larch (*L. sibirica* Ledeb.), Siberian spruce (*Picea obovata* Ledeb.), common pine (*Pinus sylvestris* L.), Siberian cedar pine (*P. sibirica* Mayr.) and Siberian fir (*Abies sibirica* Ledeb.). The Siberian cedar pine has not been tested in gas chambers. Its position in the list of ranking was determined by its resistance to sulphur dioxide and other sulphur-containing compounds including sulphur organic ones. It is likely to possess the same fluorine resistance as common pine and barely exceeds it. Such a pattern of conifer classification is typical of 20- to 40-year-old trees. In young forests, pine and spruce may rank first in resistance depending on soil conditions, presence or absence of deciduous canopy and thickness of snow cover. In old stands, pine is often ranked second in this pattern after larch.

Higher fluorine resistance of Siberian spruce as compared to pine [according to gas resistance common spruce (*Picea abies* Karst.) is usually followed by silver fir (*Abies alba* Mill.)] is accounted for by anatomo-physiological characteristics; for instance, the presence of highly developed cuticle and epidermial cell walls in needles and a lower sap pH (4.5), even compared to larch (pH 3.6).

Neighbouring species in the list do not differ significantly in their fluorine resistance; thus, the resistance of Dahurian larch differs from that of Siberian fir by less than, for example, from flat-leaved birch (*Betula platyphylla* Sukacz.), though birch is intermediate in terms of this parameter. If fluorine resistance of the black cotoneaster (*Cotoneaster melanocarpus* Lodd.), the most gas resistant among deciduous woody and bush plants tested, is taken as 10 representative points, the plane-leaved birch resistance would be 4–5 points, while the resistance of all conifers would range from 0.6 to 0.1 points (according to survival under increasing hydrogen fluoride concentration).

Different conifer species undergo the process of weakening, disease development and tree mortality in different ways. The larch is the first to respond to injurious fluorine effect. However, it can rapidly adapt by transforming its metabolism and can resist even heavy pollutant exposure for a long time. The spruce shows visible signs of injury (chlorosis and necrosis of needles) later than other conifers and its physiological weakening takes longer. Yet, the period from irreversible weakening of this species to its complete decline is the least. In pine, the extent of visible injury corresponds to the physiological state of the tree and the period of drying (starting from irreversible weakening) may be prolonged.

Not only woody plant species but also their populations are heterogeneous with respect to the extent of resistance to pollutants; differences in resistance of particular trees of a single population are also rather great. This gives an opportunity in nurseries to select plant material that is relatively resistant to industrial pollution.

The experiment described below (see Table 4) demonstrates that besides rapidly dried trees relatively resistant individuals exist, the needles of which are injured by 30–50% by 0.3–0.8 mg/m^3 fluorine. Following a 30-day fumigation period with a constant concentration of hydrogen fluoride of 0.05–0.06 mg/m^3, the trees exhibited still greater variation in their degree of injury. Among all conifers there were specimens with faint visible signs of injury and some pine and spruce trees actually showed no signs of injury, while maximum needle damage was 50% in larch, 15% in pine and 10% in spruce.

Certain structural and functional traits are responsible for the enhanced resistance of some conifers to acid gases. For instance, higher gas resistance was shown in those pine specimens which had a high water content, a high osmotic pressure of the cell sap, an enhanced content of total and protein nitrogen, poor aeration of needles and were able to control transpiration (Sokov and Gabib-Zade 1977). Somewhat higher gas resistance was observed in trees with spreading crowns, longer needles and intense annual growth of lateral shoots (Negrutskaya 1970). Spruce clones resistant to acid gases have more sensitive mechanisms for controlling stomatal opening and transpiration; their stomatal apparatus functions with greater caution towards pollutants and results in less admission and accumulation of sulphur dioxide and fluorine in needles than in non-resistant clones (Braun 1977). Higher buffer capacity and insignificant inhibition of enolase are observed in Austrian pines which are highly resistant to gases (Knabe 1975). The capacity of trees to maintain xylem and phloem growth, i.e. to renew the conducting tissues of downward and upward solute flow, is believed to be of great importance.

In conclusion, it is necessary to mention age differences with respect to conifer resistance to fluorine. Of the species studied, except for larch, trees that are 10–40 years old show somewhat greater viability than mature and old trees. These differences are particularly profound in pine but are insignificant in spruce, while in larch individuals of 80–120 years old are, as a rule, more resistant than young trees.

2.2 Factors Responsible for Duration of the Tree Decline Process

In addition to species classification and individual resistance the rate of the process of conifer decline is attributed to a number of other factors, the main ones being concentration of atmospheric pollution, duration of exposure to pollutants and environmental conditions (climate, soils and relief).

The duration and character of tree decline differ greatly under high, moderate and low fluorine concentrations in the air. Trees die particularly fast under conditions of critical injury when exposed to high concentrations of hydrogen fluoride, e.g. emissions during smelter failure. Yet a greater number of conifers in the vicinity of aluminium smelters die as a result of gradual accumulation of fluorine in their tissues; pathological processes increase until an irreversible stage is reached.

Trees in chambers fumigated continuously with very high concentrations of hydrogen fluoride (0.3–0.8 mg/m^3) have extensive damage to needles in the entire crown. Necrosis appears first in the needles of the current-year shoots and spreads to the rest of the needles in 10–15 days in larch, in 20–25 days in pine and in 30–36 days in spruce. Subsequent fumigation causes the death of buds and shoots and the decline of the whole tree. It is noteworthy that needles accumulate 1.3–1.5 times less fluorine under these conditions than during fumigation with lower concentrations of fluorine (0.05–0.06 mg/m^3) that are considered to be the minimum fluorine content in the ambient atmosphere of the zone of high fluorine concentration in the vicinity of aluminium smelters (Table 4). Actually this fluorine concentration is so high as to be lethal to all conifers even with occasional exposure, though it does not produce such drastic needle injury. This phenomenon can be accounted for by the rapid destruction of needle function under very high fluorine concentrations, resulting also in inhibition of gas exchange and transpiration; however no serious disorders in phloem flow have yet been

Table 4. Accumulation of fluorine in needles and needle damage under continuous fumigation of trees for 30–36 days with different concentrations of fluorine

Species	Control (mg% DW)[a]	HF concentration (mg/m³)			
		0.3–0.8		0.05–0.06	
		Accumulation (mg% DW)	Damage (%)	Accumulation (mg% DW)	Damage (%)
Larch	2.17 ± 0.056	13.60 ± 0.186	100	17.15 ± 0.198	50
Pine	2.50 ± 0.058	9.98 ± 0.252	90	13.41 ± 0.126	15
Spruce	2.23 ± 0.049	8.55 ± 0.201	90	12.27 ± 0.143	10

[a] mg% DW, mg fluorine per 100 g dry weight.

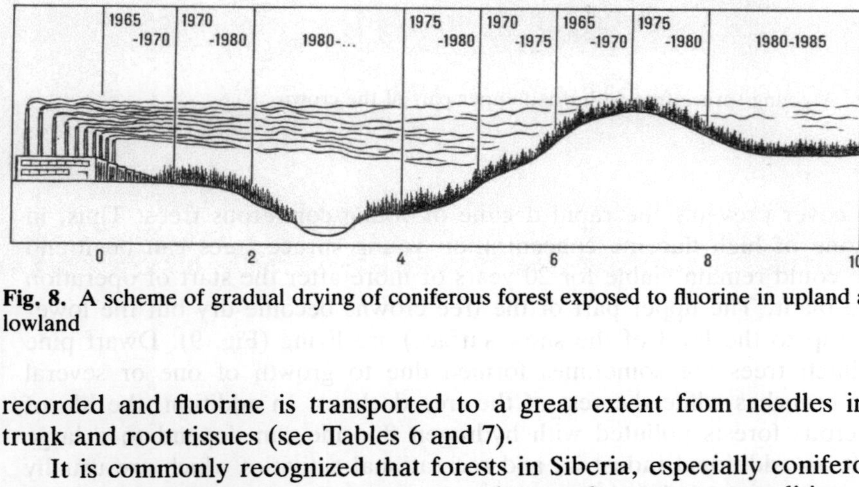

Fig. 8. A scheme of gradual drying of coniferous forest exposed to fluorine in upland and lowland

recorded and fluorine is transported to a great extent from needles into trunk and root tissues (see Tables 6 and 7).

It is commonly recognized that forests in Siberia, especially coniferous ones, are susceptible to any damage owing to the extreme conditions of growth that generally prevail. Reduced resistance to harmful impacts and a poor capacity for regeneration of woody plants and whole forest ecosystems are attributed to severe winter frosts, strong winds, vast expanses of permafrost, unfavourable seasonal changes of subsoil water and relatively poor soil fertility. Hence, coniferous forest decline resulting from industrial emissions is proceeding more rapidly here than, for instance, in central Europe where aluminium smelters have been functioning continuously since the late 1930s; only in the last 10–15 years has a final decline in adjacent coniferous forests been seen.

Relief of the land is of prime importance for localized distribution of fluorine concentrations in polluted areas. Our data indicate that trees situated 2–4 km from the smelter may decline later than those at sites as far as 8–10 km from the plant (Fig. 8). Snow, as has already been mentioned, is the main absorbent for gaseous and dusty industrial emissions in winter. It accumulates and stores within it a large amount of fluorides. Nevertheless,

Fig. 9. A young spruce tree with dried upper part of the crown

snow cover prevents the rapid decline of young coniferous trees. Thus, in the zone of high fluorine concentration young spruce trees can be found which could remain viable for 20 years or more after the start of operation of the plant. The upper part of the tree crowns become dry but the lower parts (up to the level of the snow surface) are living (Fig. 9). Dwarf pine and larch trees are sometimes formed due to growth of one or several lower branches while the rest of the tree declines. In addition, the life of coniferous forests polluted with hydrogen fluoride can depend to a large extent on additional industrial and recreational activities which are usually numerous around industrial centres.

2.3 Accumulation of Fluorine by Trees in Zones Varying in Fluorine Concentration

Assays of pollutant content in plants are in many cases a principal method of monitoring the general levels of pollution in a place, making it possible to quantify the content of pollutants in the air. Advantages of foliage (needles) analysis are particularly evident in cases where it is necessary to study the impact of long-term pollution on ground at an appreciable distance from the source of emissions. Information on total accumulation of a pollutant may likewise be used as a comparative index of gas resistance of plants in order to characterize their gas-absorbing capacities (Chuvaev et al. 1973; Getko et al. 1978). Moreover, determination of the pollutant content combined with the study of functional disorders may provide early diagnosis of injury to the plant or to the whole forest community (Keller 1977).

In the area of our study the state of stands has been examined and samples have been analysed for fluorine accumulation in zones of variable fluorine concentration around one of the aluminium plants. Mixed samples of needles collected from three to five trees during the second part of the growing season were assayed (Fig. 10). The content of fluorine in the needles of trees in the zone of high atmospheric fluorine concentration was as high as 12–20 mg% dry weight (mg per 100 g dry weight) which, in most cases, was twice as great as those in the moderate concentration zone. However, there were exceptions: in 1977 in the case of spruce, and from 1977 to 1978 in the case of larch when fluorine content was 12–13 mg% dry weight in the zone of high concentration and reached 7–9 mg% dry weight in the zone of moderate concentration.

An appreciable increase in needle fluorine content in mature and young trees in 1981 in the zone of high concentration could not be attributed

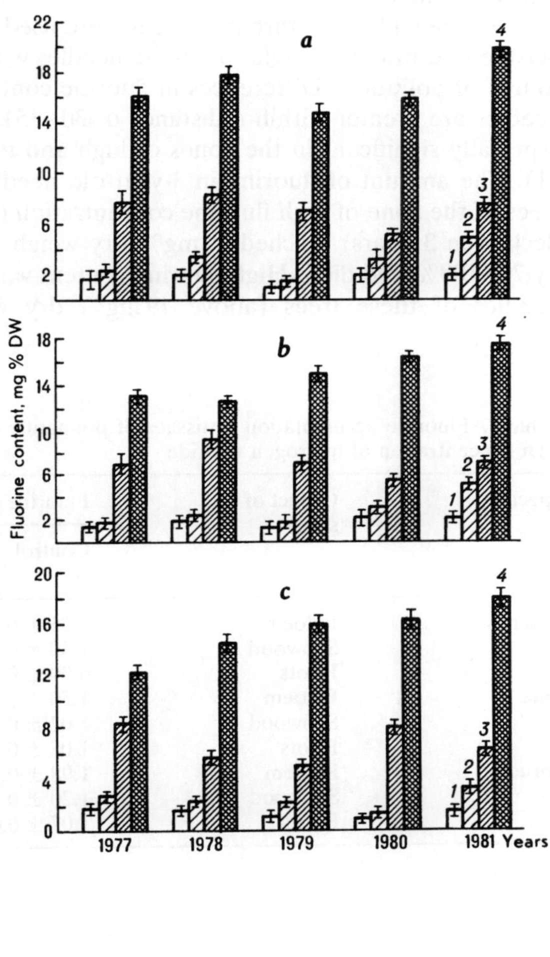

Fig. 10. Fluorine content in needles of **a** mature pine, **b** larch and **c** spruce in zones of slight (*2*), moderate (*3*) and high (*4*) fluorine concentrations (*1* indicates the control)

to changes in needle properties in severely damaged trees. A certain rise in fluorine content in the needles was also reported in the same year in the zone of low concentration. From year to year variations in fluorine accumulation could be induced by changes in the amount of emission from and performance of the smelter, by weather conditions and consequently, by the rise or fall of fluorine concentration in the atmosphere. Fluorine content was also high in the needles of all species in the zone of moderate concentration. Trees that had been irreversibly weakened underwent extensive drying in this zone. The decline followed the spread of needle necrosis throughout the crown with the appearance of dry-topped trees.

In the zone of low concentrations the level of fluorine accumulation in the needles was also above that of the control. The needles of young trees were not visibly damaged whereas mature pine needles showed slight necrosis (up to 10%). Conifers in this zone may suffer from a cumulative effect of fluorine. In the zone of high concentration the content of fluorine (mg% dry weight) in the phloem and sapwood of the trunks and in the root tissues of mature and young trees was significantly above the content of controls (Table 5).

Mature and premature pine trees were used to elucidate the relationship between fluorine accumulation in the needles with age and distance from the source of pollution. Differences in fluorine content in 1 and 2- to 3-year-old needles are greater within a distance of 30–35 km from the smelter and are especially significant in the zones of high and moderate concentration (Fig. 11). The amount of fluorine in 3-year-old needles of irreversibly weakened trees of the zone of high fluorine concentration (the majority of which would decline in 3 years) reached 29 mg% dry weight; the needles were damaged by 70–100% and died. High fluorine content was also observed in 2-year-old needles of these trees (above 19 mg% dry weight). Young needles on

Table 5. Fluorine accumulation in tissues of premature and mature trees in the zone of high concentration of hydrogen fluoride

Species	Object of study	Fluorine content (mg% DW)	
		Control	Zone of high concentration
Larch	Phloem	1.25 ± 0.076	3.93 ± 0.176
	Sapwood	1.30 ± 0.076	4.12 ± 0.178
	Roots	0.90 ± 0.086	1.73 ± 0.061
Pine	Phloem	1.50 ± 0.058	4.45 ± 0.120
	Sapwood	1.02 ± 0.083	2.87 ± 0.173
	Roots	1.08 ± 0.087	1.52 ± 0.098
Spruce	Phloem	1.98 ± 0.119	7.37 ± 0.084
	Sapwood	1.20 ± 0.106	4.40 ± 0.177
	Roots	1.07 ± 0.049	1.18 ± 0.075

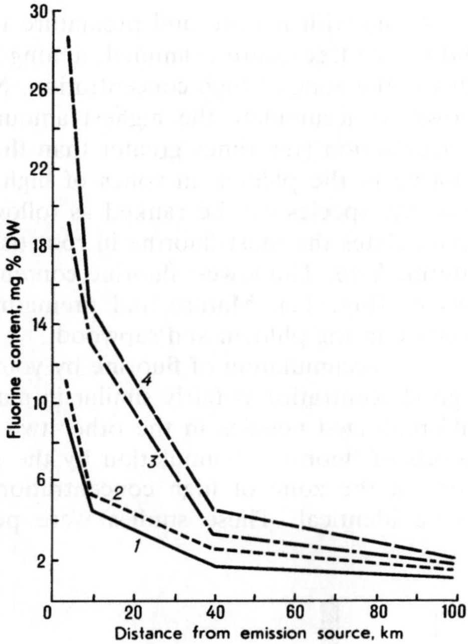

Fig. 11. Fluorine content in needles of (*1*) current-year shoots, (*2*) 1-year-old, (*3*) 2-year-old and (*4*) 3-year-old needles of mature pine trees situated at different distances from the emission source

current-year shoots did not show visible signs of injury at the time of sampling, i.e. in the middle of the growing season, within 15 km of the smelter; they were damaged by 20–30% within 4 km and had undergone 40–50% necrosis on average with a fluorine content of about 9 mg% dry weight within 1.5 km.

In the zone of high fluorine concentration it does not take long for young needles of current-year shoots to accumlate 8–9 mg% fluorine of dry weight and to show visible signs of injury. As rapidly as by the mid-growing season the upper third of single needles may be necrotic. Rapid appearance of burns in young needles may be explained not only by their enhanced sensitivity to atmospheric pollutants but also by the flow of fluorides along with metabolites into developing needles from storage tissues. The possibility of flow of fluorides from storage tissues to young leaves in spring is shown by Keller (1974a, 1975b) in his experiments with conifers and deciduous trees.

In the zone of low concentration of fluorine the state of stands has essentially not changed during the period of observation (1977–1981). The trees here failed to show signs of necrosis or were damaged only slightly (not more than 10% needle damage). Consequently, the level of fluorine accumulation in the needles was not high (see Fig. 10). The data obtained on 2- to 2.5-fold increase in fluorine content as compared to the control were an exception, indicating potential high risk for conifers in this zone.

Along with mature and premature trees 12- to 15-year-old larch, pine and spruce trees were examined, among them those that survived in several sites in the zone of high concentration. Needles of young spruce trees were shown to accumulate the highest amount of fluorine in the zone of high concentration (ten times greater than the control). In terms of decrease of fluorine in the phloem in zones of high and moderate fluorine concentration, the species can be ranked as follows: larch, spruce and pine. Spruce accumulates the most fluorine in root tissues, larch the least and pine is an intermediate. The lowest fluorine content is observed in the sapwood of all species (Fig. 12). Mature and premature trees have a differing ratio of fluorine in the phloem and sapwood.

The accumulation of fluorine by young and mature trees in the zone of high concentration is fairly similar in magnitude. Fluorine accumulation by different-aged needles in the other two zones also shows little difference. Levels of fluorine accumulation by the needles of young and mature pine trees in the zone of high concentration of fluorine emission were found to be identical. These studies were performed in July and August for

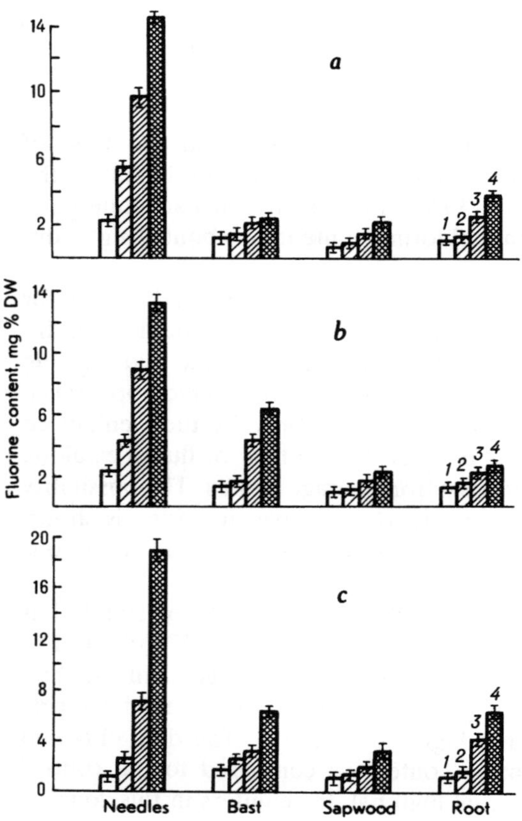

Fig. 12. Fluorine content in **a** young pine, **b** larch and **c** spruce in zones of slight (*2*), moderate (*3*) and high (*4*) fluorine concentration (*1* indicates the control)

three consecutive years. The average content of fluorine in the needles of mature and young trees was 18.87 and 19.39 mg% dry weight respectively. Threshold levels of fluorine accumulation in the needles have been determined; these show little or no dependence on tree age. In the case of irreversibly weakened pine trees in the zone of high concentration of fluorine emission, the threshold level of fluorine accumulation in those needles still living (i.e. with some chlorophyll) is about 20 mg% dry weight. The same threshold of fluorine accumulation is recorded for the needles of young pine trees in fumigation chambers that are supplied with a daily fluorine dose of 0.2–0.3 mg/m^3 in air during the whole of the growth period.

In totally yellow needles, just before they fall from the tree, the fluorine content may increase up to 29–31 mg% dry weight. By our observations this is the threshold level of pollutant accumulation in dead needles. In experiments where bases of pine branches were immersed in sodium fluoride (0.02%) the needles lost chlorophyll in 20–21 days and had the same fluorine content of 28–31 mg% dry weight just before loss. In cases of a drastic increase in atmospheric fluorine concentration resulting from the occasional escape of gas wastes through factory ventilation systems, the needles can die with an even lower content of fluorine in them (14 mg% dry weight, or less). Pine needles fumigated in chambers with 0.3–0.8 mg/m^3 fluorine accumulated 9–11 mg% dry weight by the time of their loss. This indicates that an atmospheric fluorine concentration above 0.3 mg/m^3 is highly lethal and under such conditions needles die without even having time to accumulate the threshold amount of fluorine.

2.4 Changes in Chemical Composition of Pine Needles in the Zone of Emissions from Aluminium Smelters

The chemical composition of pine needles, as a representative forest species, was analysed from trees in the area of emissions from two large-scale aluminium smelters. This section presents the results of the effects on the chemical balance by emissions from the higher powered of these two smelters, the Bratsk aluminium plant. The town of Bratsk forms one of the large-scale industrial centres in the eastern USSR with the aluminium smelter and a large lumber industry providing the main causes of environmental pollution. A very large area around the town is polluted and the degree of toxic emissions is unacceptably high. Fluorides, sulphur-containing compounds, chlorine, nitrogen oxides and carbon monoxide are among the chief pollutants. The harmful impact of the Bratsk aluminium smelter on forest vegetation seems to be greater than of any other aluminium smelter in the country. The heavy impact on forests is determined, in addition to the contribution by inadequate purification systems, by a highly broken landscape, the presence of a large number of windbroken slopes, the close proximity of a reservoir that causes numerous local air currents, enhanced air moisture, acidic rainfall, dews and fogs.

The coniferous forests in this region have been under examination since the 1970s when most attention was paid to the phytotoxic effect of fluorides as the most aggressive pollutants. Comprehensive studies of the effect of other pollutants and their combined effects on the forests have been underway since 1987. Here data will be reported on changes in chemical composition of needles of pine trees growing under different conditions and within different distances from the Bratsk aluminium smelter.

Maximum permissible concentrations for the safety of public health of harmful atmospheric agents in the locality of the plant were specified in the design of the Bratsk aluminium smelter. However, these standards were not met and were also set at too high an original level. As a result, there has been an extensive decline in forest; this was widely discussed in the press. The first dry trees to be affected by smelter emissions appeared in 1968. Since then regular examination of the forest has shown a constant increase in its decline with additional contribution by pollutants from other industries. By 1985 the total area of stands damaged to a variable extent in the polluted location was over 80 000 ha. This value does not include dried trees that had to be felled but would have comprised the main part of the green zone of the town.

During the initial period of air pollution rapid tree decline, both in the town and in adjacent forests, was caused by critical toxic exposure. At distances farther from the source of emission the onset of tree decline occurred much later and resulted from the cumulative effect of pollutants (basically fluorine) and from gradual loss of resistance under conditions of the chronic toxic effect. At present this is the main type of forest decline in the region. In the near future the zone of stand decline is expected to continue spreading in this way without a halt in the process. According to Pleshanov (pers. comm. 1985), at the Siberian Institute of Plant Physiology and Biochemistry in Irkutsk, the area of decline of forest exposed to industrial emissions from Bratsk could be expected to almost double by the year 2000 if pollution remains at the same level.

Intensive monitoring of accumulation of pollutants in indicator plants is necessary under such circumstances of grave ecological imbalance. Evergreen species of conifers are the most suitable for this study and in the region considered the common pine was chosen as a representative forest species. To examine stands we used chemical analysis of assimilatory organs as the best indicator of the trees' state. Numerous data on foliage analysis show not only toxic substance accumulation under the impact of pollutants but also total changes in the chemical composition of needles (Landolt and Scherrer 1985; Reigber and Brawn 1985; Landmann et al. 1987; Raunemaa et al. 1987; Klaus 1989). Analysis of the balance of chemical elements provides a very reliable assessment of the physiological state of plant organisms and, in combination with visual examinations, allows both particular stands and whole forest regions to be diagnosed.

In addition to fluorine and sulphur accumulation we followed changes in the content of heavy metals present in industrial emissions; these include lead, cadmium, copper and those accompanying fluorine emission such as aluminium and silicon. We also determined the content of nitrogen, phosphorus, potassium, calcium, magnesium, sodium and manganese in the needles to estimate the physiological condition of the tree. Two-year-old needles from the whole crown were sampled from five 30- to 40-year-old pine trees at different distances from the Bratsk aluminium plant. Trees of the same age from a location at more than 100 km from populated industrial regions were used as a control.

The data obtained (Table 6) show a significant emission load on the area. Disturbances in the physiological state of woody plants are shown not only by the enhanced level of toxic elements in the needles but also by changes in the content of a number of principal macroelements and a decline in manganese level. Manganese is a very important microelement in plants since it is involved in all basic metabolic processes. Manganese deficiency causes a reduction in the intensity of photosynthesis, disturbances in respiration and disturbances in metabolism of nitrogen, nucleic acids and auxin. The oxidative properties of manganese give it a specific role in maintaining essential reduction-oxidation conditions in cells. Therefore, a low content of manganese in the needles indicates the harmful effect of emissions on tree physiology as far as 40 km from the aluminium plant.

The concentration of fluorine in the needles of trees close to the smelter approaches the level characteristic of trees in a state of irreversible weaken-

Table 6. Changes in chemical composition of pine needles exposed to emissions from the Bratsk aluminium smelter (% DW with exceptions indicated)

Element	Distance from smelter (km)				Control
	1.5	3.0	10.0	40.0	
Ca	0.479	0.499	0.591	0.524	0.996
Mg	0.188	0.147	0.190	0.144	0.248
K	0.48	0.40	0.37	0.48	0.86
N	1.37	1.36	1.24	1.14	1.21
P	0.13	0.13	0.13	0.11	0.15
Na	0.014	0.020	0.018	0.017	0.023
Mn	0.007	0.027	0.008	0.012	0.071
Al	0.067	0.121	0.078	0.031	0.056
Si	0.111	0.150	0.219	0.042	0.041
S	0.09	0.09	0.12	0.08	0.05
Cu (mg kg^{-1})	3.7	3.6	3.1	3.2	2.8
Pb (mg kg^{-1})	1.7	2.3	1.1	1.9	0.6
Cd (mg kg^{-1})	0.021	0.028	0.005	0.009	0.009
F (mg% DW)	18.7	15.2	14.0	3.9	0.9

ing. The content of this pollutant in the needles is still rather high even at 10 km from the source of emission. The content of sulphur in the needles is one of the principal diagnostic parameters; however, some caution should be exercised in using it as such. Although sulphur analysis is considered to be acceptable for determination of plant state and extent of air pollution, it should be used very carefully because, under normal conditions, its content varies greatly in plants. Therefore, to estimate the extent of emission load it is insufficient to use only total sulphur content as an indicator; simultaneous determination of concentrations of other elements including basic compounds should be used. The evidence demonstrated that sulphur content was high in all samples collected, regardless of distance from the smelter.

Aluminium and silicon levels were assayed for possible use as markers of fluorine emission. The main part of the solid aerosol fraction present in emissions from aluminium smelters is composed of these elements. Silicon is also present in the form of a gaseous tetrafluoride that readily enters leaf tissues through stomata. An appreciable rise in the content of these elements is observed at a distance of 10 km from the plant, while at greater distances the concentration shows an obvious decrease. High accumulation of these elements in the needles of pine trees in the vicinity of the plant suggests profound disorders in the physiological processes of the trees. An excess of aluminium is known to affect phosphorus metabolism in plants. Aluminium precipitates a large part of the phosphorus in cells resulting in a phosphorus deficiency for metabolism. Aluminium absorbed by cell walls may rapidly block active sites and so upset ion exchange significantly (Salyaev 1969).

Silicon as a pollutant has attracted little research interest before; it was mainly the negative effect of silicon deficiency on metabolism that had been studied. It was not until recently that information appeared considering another aspect: namely, the effect of excess silicon on the plant organism. It has been shown that the level of silicon in needles serves as an indicator of various stresses, among them the impact of pollutants (Knabe et al. 1989). In particular, a positive correlation was found between silicon and fluorine concentrations in needles. It was concluded that tree viability declined with a high silicon content in needles.

Stands in the region of aluminium smelter emissions are further aggravated by heavy metal accumulation. Mechanisms of the phytotoxic effect of heavy metals are diverse. Bowen (1966) presents the following classification of these mechanisms: (1) disturbances in enzyme function; (2) formation of antimetabolities; (3) chelation of the main metabolities; (4) acting as catalysers of decomposition of main metabolities; (5) changes in permeability of cell membranes; and (6) substitution of structural elements in the cell.

We have assayed copper, lead and cadmium in the needles. Copper did not exceed by much the control level and therefore could not provide a negative effect. According to some authors (Duda 1981; Lyanguzova and

Chertov 1990) copper concentrations must exceed by 3- to 30-fold the control level to produce a phytotoxic effect. Lead and cadmium contents were shown to be increased. An enhanced level of lead accumulation was observed within 40 km of the plant. The content of cadmium was two to three times greater than that of the control. Among heavy metals cadmium and lead are the most toxic for plants (Glattes 1985; Godbold et al. 1985). They induce a considerable decrease in the content of potassium, calcium, iron, zinc and manganese in leaves; their joint effect causes greater inhibition of plant growth than the effect of each element singly (Kahle et al. 1989). In the surveyed stands that were exposed to emissions from the smelter high levels of lead and cadmium and absorbed gaseous pollutants in the needles were accompanied by a decrease in the content of potassium, magnesium and calcium.

Biochemical assays were supplemented by visual examination of tree damage. The stands were damaged up to 40 km from the plant and within this distance the zone of irreversibly weakened and drying stands could be seen to extend as far as 10 km from the aluminium smelter. By 1989 the forest had declined on the upland of this zone and surviving conifers were characterized by high accumulation of fluorides, sulphur, silicon, aluminium, lead and reduced levels of basic macroelements and manganese. Such trees are bound to die. The trees at a distance of 40 km have low needle injury. Nevertheless, the long-term effect of lower concentrations of toxic emissions may be concluded on the basis of increased levels of fluorides, sulphur and lead in needles and an obvious deficiency of the main macroelements (with the exception of nitrogen) and manganese. The stands are chronically weakened in this zone and if the present emission load remains the same the damage to trees would continue.

2.5 Dynamics of Seasonal Fluorine Accumulation and Injury of Trees

The study of dynamics of seasonal fluorine accumulation in pine needles in the zone of impact of aluminium plant emissions shows mature and young trees to have a similar pattern of change (Fig. 13). The largest amount of fluorine in the needles is observed in winter. Its content has decreased by the beginning of the growing season, is minimal from May to July and increases again at the end of the growing season. This pattern of dynamics of fluorine accumulation is regulated primarily by seasonal physiological changes in plants. Maintenance of gas exchange (though very low) on the one hand and cessation of growth processes with relative stabilization of needle dry weight during winter on the other hand, facilitates fluorine accumulation in the needles. In spring and summer, following qualitative

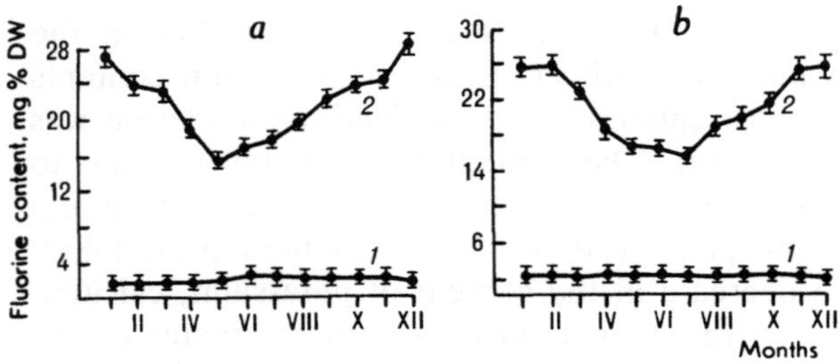

Fig. 13. Dynamics of seasonal fluorine content in needles of **a** young and **b** mature pine trees in non-polluted stands (*1*) and in the zone of high fluorine pollution (*2*)

changes in metabolism, fluorine is translocated together with metabolites from the needles to trunk and root tissues. A part of it is removed from the needles by transpiration. In addition, leaching of fluorine by summer rainfall can lead to a certain decline in toxic elements in the needles and this should not be overlooked.

Concurrent visual observations of the changes in extent of needle damage to the same trees corroborates the above pattern of fluorine changes. The needles are most heavily damaged from November until March. The percentage of necrotic needles declines in late March-April due to the loss mainly of 3-year-old and partially of 2-year-old needles that are completely dried. With the appearance of young shoots covered with new green needles the amount of damaged needles in the crown decreases in May–July. In the second half of the growing season needles of the past year are increasingly damaged and young needles exhibit visible signs of injury with as much as 20–50% necrosis by as early as October in the zone of high fluorine concentration. In winter months and in March approximately one-half of 1-year-old and one-third of 2-year-old needles remain green. It is noteworthy that young trees have more pronounced critical stages of maximal and minimal accumulation of fluorine; this seems to correspond to age characteristics of the dynamics of metabolic processes.

The degree of plant damage and severity of their subsequent toxic state depends largely on the timing of fluorine exposure. To determine the period of greatest hazard of fluorine to conifers 12- to 15-year-old pine, larch and spruce trees were artificially fumigated with 0.2–0.3 mg/m^3 hydrogen fluoride at different periods of growth. The trees were fumigated with fluorine in large chambers from May to the beginning of June, in July and in September over 20 consecutive days in each period. The damage was assessed by visual examination of needle injury in the year of fumigation and during subsequent years, and by determining fluoride accumulation in the needles. The fluorine content in the control is an average of the results of analysis of healthy needles at different stages of growth (Table 7).

Larch was the first to suffer damage in the chambers as distinguished by thinner wall epidermis, relatively thin cuticle and insufficiently devel-

Table 7. The content of fluorine (mg% DW) in the needles of trees fumigated with hydrogen fluoride

Period of growing season	Current year	After 1 year	After 2 years
Larch			
Beginning	13.37 ± 0.219	7.20 ± 0.198	3.47 ± 0.099
Middle	15.38 ± 0.206	3.55 ± 0.180	3.53 ± 0.178
End	8.68 ± 0.151	7.78 ± 0.164	1.80 ± 0.148
Control	1.83 ± 0.061	2.02 ± 0.095	1.53 ± 0.102
Pine			
Beginning	20.62 ± 0.270	14.93 ± 0.241	3.02 ± 0.182
Middle	9.48 ± 0.202	8.10 ± 0.309	6.33 ± 0.109
End	7.48 ± 0.246	5.43 ± 0.282	4.35 ± 0.257
Control	2.33 ± 0.088	2.47 ± 0.161	2.20 ± 0.115
Spruce			
Beginning	8.72 ± 0.198	5.97 ± 0.099	2.37 ± 0.143
Middle	10.32 ± 0.206	2.52 ± 0.180	2.48 ± 0.140
End	16.12 ± 0.185	2.95 ± 0.143	2.40 ± 0.093
Control	2.25 ± 0.089	2.32 ± 0.070	2.22 ± 0.091

oped hypodermis compared to other coninfers. With exposure to hydrogen fluoride at the beginning of the growing season during the period of intense growth and shoot formation its needles were damaged almost totally and accumulated a great amount of fluorine. In the two weeks following the fumigation new needles appearing on brachyblasts showed distinct chlorosis and were of a smaller size than normal ones. In the middle of the growing season when shoots had completed their growth in length, the needles were totally damaged by the gas and had a high fluorine content. The harmful effect of unfavourable defoliating factors in this period is associated with the loss of the ability to produce new foliage.

In these experiments exposure to hydrogen fluoride at the end of the growing season resulted in 30–40% damage to needles. The amount of fluorine accumulated during this period was accordingly less; this seems to be related to the diminished physiological activity of autumn needles, thickening of the ground tissue and intense efflux of fluorine along with assimilates prior to defoliation. The following year the fluorine content of the needles of all larch trees increased significantly above the control level. The plants fumigated at the onset of growth suffered the greatest damage. Their needles grew only in the lower part of the crown and the amount decreased further in the third year. The needles of the plants fumigated in the mid-growing season failed to develop during the second year except on the main and some lateral shoots. Needle size both in auxiblasts and brachyblasts (hypertrophic, compensating needles) significantly exceeded the control but after a month signs of chlorosis appeared and increased. In the two years following the fumigation the size of needles of these trees did not differ from those of the control. The needles of the trees fumigated at the end of the growing season were not visibly damaged the following year, though auxiblasts were 1.5 to 2.0 times shorter and in the third year only some last-year shoots lost their needles.

In pine and spruce the pattern of damage was different. Visible damage of needles of both species was comparatively greater during spring fumigation than during fumigation in the middle and at the end of the growing season. However, in the second case the needles accumulated a high amount of fluorine (see Table 7). The effects were profound but were exhibited later. In pine there was an intense yellowing and partial loss of the needles, in spruce the injured needles were lost and in both species a greater part of the apical and lateral buds died and there was no or only poor growth of new shoots. The very short shoots that formed from the few dormant buds of damaged pine trees had shorter or hypertrophic needles which became yellow and dried up later. The appearance of these trees did not change much in the two years following the fumigation.

In spruce particularly severe damage by fluorine was observed following fumigation at the end of the growing season. Most of the needles of trees weakened by fluorine during preparation for the winter season dried and were lost during the next spring. The comparatively low fluorine content in spruce needles in the year following the fumigation is accounted for by the fact that only needles of new shoots were analysed as the rest were lost. The rather intense process of needle formation was followed over 2 years on new shoots of spruce trees fumigated at the beginning and in the middle of the growing season.

2.6 Reduction in Annual Wood Growth and Extent of Xylem Flow with Tree Decline

Air pollution induced by industrial emissions is one of the principal reasons for reduction in forest productivity. A decrease in the rate of wood production induced by acid pollutants has been mainly illustrated with sulphur dioxide. Few data are available on the effect of fluorine-containing emissions. The experiments in 1971–1974 on increment dynamics in conifers affected by aluminium plant emissions appear to be the most comprehensive (Sokov and Rozhkov 1975).

In this work, the increment was analysed in young and mature pine, spruce and larch trees growing in zones of high, moderate and low fluorine concentrations. The width of annual rings was calculated as an arithmetic mean of four measurements for each cross-section (north, south, east and west) and of 16 measurements for these cross-sections along the trunk. Six standard trees of each species were chosen as a model for control analysis at particular sites in each polluted zone.

Table 8. Decrease in the average radial increment of the trunk in mature trees in zones with varying fluorine concentration after 5 years after the start of plant operation

Species	Fluorine concentration	Annual increment[a] (% of many years' mean)
Pine	High	72.0 ± 5.2
	Low	88.1 ± 4.5
Spruce	High	48.2 ± 16.0
	Intermediate	85.0 ± 9.2
	Low	93.0 ± 5.6
Larch	High	66.3 ± 15.7
	Intermediate	87.0 ± 6.6
	Low	91.0 ± 8.3

[a] An average increment over several years (many years' mean) is taken as 100%. An annual increment is expressed as a percentage of this value.

The results obtained, expressed as a percentage of the mean value of annual increment during the 13 years prior to the start of smelter operation (taken as 100% and termed many years' mean), illustrate changes in growth both for an average trunk and at various heights along the trunk. The annual radial increment for an average trunk 5 years after operation of the plant began is as shown in Table 8.

Under conditions of continuous pollution the continued production of annual wood is an essential part of the defence mechanism of the weakened tree; it allows for regular removal of fluorine from the conducting tissue of xylem and its renewal. It is evident (see Table 8) that the reduction in radial increment ranges from insignificant to intermediate values (28–52%) in weakened but still viable trees of all three conifer species following a 5-year exposure to the toxic effects of fluorine in the zone of high concentration. This range in reduction is not high for severely damaged trees in this zone. In the zone of intermediate concentration of fluorine the increment reduction was minimal (13–15%) and in the zone of low concentration no statistically significant reduction in increment was seen. A surprisingly larger reduction in the radial increment was revealed in larch and spruce as compared to the less resistant pine. Further research is necessary to explain this phenomenon. For the present it may be assumed that different species of conifers differ in their ratio of intensity of regeneration processes between the trunk and crown.

The growth of annual wood in weakened trees (excluding drying trees) was no less than 30% of many years' mean after 20 years of smelter operation in this area (unpublished data of Voronin 1986). Thus, one-quarter of all premature and mature trees of pine forests lying outside of the zone of moderate fluorine concentration displayed only an occasional reduction in radial increment (slightly damaged trees); the next quarter of trees showed 40–55% reduction of many years' mean (weakened trees);

more than one-third of trees exhibited 56–70% reduction (chronically sick trees); and only in one-seventh of trees did the reduction exceed 70% (declining trees). Interesting variations in radial increment were followed by Voronin in premature and mature spruce trees that survived in the low-lying riverside land in the zone of high fluorine concentration. Within the first 9 years their wood production decreased by 43% on average (by 31% in some trees) and within the next 11 years it increased and reached 50.5% (with a maximum of 69%) of many years' mean. Pine trees, 60–80 years old, and larch also retained their resistance to fluorine exposure in this low-lying land.

Sterba (1984) attempted to estimate comparatively the effect of pollution and phyllophagous insects on conifer stands. Spruce was the subject of his studies and he found that emissions over many years from the aluminium plant and outbreaks of fir sawfly (*Pristiphora abietina* Christ.) were deleterious factors. The increase in height of trees and production of wood were used as the main parameters for calculation. The author inferred that fluorine affected both of these parameters whilst the injury caused by sawfly larvae exerted an influence on only the first parameter. However, this is not always the case. Particular studies performed by workers in our laboratory (Pindyura and Pleshanov 1975; Pindyura and Stepanchuk 1978 and others) demonstrated that wood production in conifers was reduced as a result of defoliation by insects in some cases but an even greater effect was caused by the impact of industry. As in the case of needle loss following exposure to pollutants the reduction in wood production of trees damaged by phyllophagous insects depends on tree species, tree age, original physiological state, environmental conditions, and timing, extent and frequency of defoliation.

Upward flow in xylem occurs along thin-walled tracheids which are responsible for early-wood formation in annual rings. Late-wood consists of thick-walled tracheids which serve as a support. The importance of preserving early-wood in the damaged plant in an amount sufficient to maintain water exchange becomes evident if one considers that the process of decline in most trees of coniferous forests exposed to pollution proceeds from the top to the bottom.

The damage of assimilation area causes not only a reduction in radial increment but also alterations in the ratio of early-wood : late-wood. These alterations have been most commonly studied in forests damaged by xylophagous insects but have been followed at length in trees artificially defoliated to a varying degree and at different times. An examination of the effects of artificial defoliation of conifers (Pindyura and Stepanchuk 1978) allows the following conclusions to be drawn. In pine and in Siberian cedar pine the thickness of the tree ring is reduced owing to decline of both early- and late-wood irrespective of the time of damage. In spruce and in larch defoliation in spring causes a reduction in growth mainly because of a decrease in the rows of thin-walled tracheids but damage to spruce in summer and autumn and defoliation of larch in summer causes a decrease in production of early- and late-wood.

The reduction in the increment of early-wood in conifers suffering environmental pollution can be noted as early as in the first stages of tree weakening. Analysis of increment in silver fir revealed that some 50- to 120-year-old trees that appeared healthy were actually sick. Their decline was followed by observing the reduction in radial increment, mainly of water-conducting wood in the lower trunk (Kiennen and Schuck 1983).

In annual tree rings of conifers exposed to hydrogen fluoride the number of thin-walled tracheids decreases throughout the trunk while the ratio of early-wood: late wood varies with tree height and depends on the extent of tree damage. For instance, shortly before the tree enters the stage of irreversible weakening the increment of early-wood may be minimal or slightly higher in the region above the butt of the trunk. As a result, the sapwood of the butt becomes highly water saturated. This process is assissted by the roots which abruptly increase the supply of water into the trunk after considerable needle loss when there is a corresponding reduction in transpiration. It should be noted that damaged needles retain a certain photosynthetic activity even when only relatively small pigmented spots remain. Only totally yellow needles with a maximum amount of accumulated fluorine are lost. The life of partially damaged needles barely exceeds 2 years. It is young current-year needles that basically maintain the minimum level of photosynthesis necessary for the tree to live.

In the zone of high fluorine concentrations around large-scale aluminium smelters (emitting up to 2500 tons of fluorine per year) the initial decline in coniferous trees appears in the third year. It usually takes whole stands a decade to decline. In the polluted environs of smelters of two to three times lower capacity conifers start declining 10–14 years after fluorine impact. The decline of trees in the zones of moderate and low fluorine concentrations lasts many years. The trees are declining for 20–30 years or more. Individual differences in fluorine resistance and particularly in the type of crown dieback are most clearly exhibited in these cases. Along with distribution of necrotic mottling in the crown from its top to the base and with gradual enhancement of injury (decline from the tree top), some plants were observed to experience complete drying of certain branches, not necessarily from the top, while the rest remained green. Such trees preserved their viability for a longer period than did those which immediately developed dried tops.

2.7 Summary

Conifer resistance to fluorine, expressed as a point system according to their survival under conditions of increasing fluorine concentration, is an order of magnitude lower than for gas-resistant deciduous trees and bushes. Among conifers of Siberia, Dahurian larch is the most fluorine-resistant species followed by Siberian larch, Siberian spruce, common pine, Siberian cedar pine and Siberian fir in order of decreasing resistance. Coniferous individuals vary considerably in their susceptibility to damage by fluorine, both under different fluorine concentrations and under different conditions of fumigation; the greatest variations occur under a prolonged effect of low concentrations. Atmospheric fluorine concentration above 0.1 mg/m^3 contributes to comparatively rapid (3–10 years) development of the process of irreversible weakening and decline of conifers. Lower concentrations are responsible for chronic injuries which can last as long as several decades.

Needle resistance to the injurious effect of fluorine and whole-tree resistance are not adequate parameters by themselves, since needle resistance is mainly determined by the species, age and anatomo-physiological behaviour of the needles and tree resistance is affected by defence mechanisms of the whole plant organism and its ability to regenerate. The level of fluorine accumulation in needles does not depend on tree age, though it rises with needle age. A threshold accumulation of fluorine in pine needles during the lifetime of the tree is about 20 mg% dry weight and the largest amount of fluorine (up to 31 mg% dry weight) is accumulated in needles during their dieback.

Fluorine accumulation in the phloem and sapwood of the trunk and in root tissues varies with age: young trees of all species studied had a greater content of fluorine in the roots and a somewhat lower content in the sapwood, unlike severely damaged mature trees where the intensity of fluorine accumulation in the phloem increases with age in pine, decreases in larch and shows no detectable variation in spruce.

The harmful effect of hydrogen fluoride on conifers was observed over the whole growth period and in winter. In larch the most severe needle damage and highest fluorine accumulation was found at the beginning and in the middle of the growing season and in pine and spruce during spring fumigation. Dynamics of average seasonal fluorine content in needles was somewhat different: the maximum amount of fluorine was recorded from November to March, it declined in March–April due to loss of dead, fluorine-saturated needles and remained low in May–July owing to the development of new shoots covered with young needles.

Fumigation with hydrogen fluoride results in prolonged after-effects as shown by an increased content of fluorine in the needles during the period some years after fumigation and by the possible appearance of visual signs of damage in species with needles which are many years old.

It is very important for the tree experiencing long-term, low concentrations of pollutants to maintain a sufficient rate of early-wood growth and an upward flow from the root system to act as defence mechanisms. Radial increment in all the coniferous species studied, which had been reduced by fluorine impact before the tree passed to the stage of irreversible weakening, did not exceed 70% of many years' mean. The main defence mechanisms of conifers suffering from a cumulative fluorine effect (in zones of low and moderate fluorine concentration) involve gradual changes in metabolism aimed at prolonging survival under conditions of extremely low levels of photosynthesis, transpiration, gas exchange and root nutrition. Such trees retain their viability following damage of 80–100% to needles from previous years and of 40% to current-year needles with living needles having a fluorine content of 18–20 mg% dry weight in July–August. Moreover, 80- to 100-year-old larch and pine trees retaining three to four branches with poor needles may be kept alive for some years. Conifer decline in stands polluted with fluorine-containing emissions mainly follows a pattern of conifer drying starting from the tree top.

Chapter 3

Fluorine Toxicosis in Conifers

3.1 The Corrosive Burn and Toxicosis in Plants

A survey of the evidence obtained in recent decades for a wide range of pollutant effects on plants permits a distinction between four types of damage which differ in the character and intensity of metabolic disorders: acute, drop-like burns, cumulative and insidious (invisible, latent). It is impossible to clearly distinguish between the groups.

Insidious injuries are induced by the effect of very low pollutant concentrations during continuous or recurrent periods. It is the name of this damage that implies the absence of visible signs of plant damage. This type of damage includes disturbances in physiologo-biochemical processes (increasing membrane permeability, alterations in the activity of a number of enzymes, photosynthetic inhibition, etc.) which can be registered only by particular methods. The extent of insidious injury depends on the type of pollutants with hydrogen fluoride causing the widest range of damage. Accumulation of insidious injuries results in decline of growth and productivity in trees and loss of resistance to climatic factors, pathogenic microorganisms and insects. Since other factors can produce similar effects, for instance, nutrient element deficiency or poor water supply, insidious injuries induced by pollutants are difficult to identify. The latent injuries are reversible; they do not result in steady metabolic disorders and after exposure to pollutants has ceased the plants recover their normal performance.

Cumulative damage appears following a long-term effect of low concentrations of pollutants. It is characterized by disorders in the main metabolic processes which leads to visible decrease in size and weight of leaves (needles), their premature loss, thinning out and drying of the top of the crown, teratological changes of plant habit, decline in fruit production, growth suppression and early ageing. The cumulative damage results in excess accumulation of fluorine in assimilatory organs and the appearance of necrosis.

Acute injuries are caused by the effect of high concentrations of pollutants. Such concentrations can occur during failure of the smelter with considerable amounts of gaseous waste escaping through the ventilation system. Acute damage primarily affects assimilatory organs and progresses over several hours (days), resulting in chlorosis and subsequent leaf tissue necrotization or death of the whole shoot. Under high concentrations of pollution not only young shoots but also 1- and 2-year-old shoots may decline rather rapidly. However, this can be referred to as a type of cumulative damage.

In recent papers on the effects of pollutants on plants the terms "burn" and "damage" ("injury") are used synonymously, but this does not seem reasonable for the case of woody plants where corrosive burn is a particular type of damage differing from other toxic injuries by its mechanism of formation, rate of development and efficiency of effect.

Drop-like burns, in contrast to the acute damage which embraces the whole needle and expands throughout the whole or greater part of the crown, look like necrotic points or larger mottled areas that cross the needle and gradually expand. In the majority of healthy needles with a highly developed cuticle, the post-burn necrosis starts in the apical region and then expands towards the base.

The acute damage of larch needles by emissions with a high concentration of hydrogen fluoride was observed in early June in young and middle-aged trees growing within 3 km of the aluminium smelter. The needles looked wilted and were light greenish-grey by the third day and began turning brown by the fifth and sixth days. Injured needles of larch rapidly dry and are lost. They are regenerated by stimulating development of next-year buds in brachyblasts. Potential next-year needles develop in the brachyblast vegetative cone at the same time as current-year needles and if the latter are lost these secondary needles appear in the same summer. Acutely damaged needles of spruce also decline rapidly, while in pine branches they remain alive for a long time. When 2-year-old pine shoots retain their viability in spite of needle damage, tight groups of short auxiblasts develop from dormant buds at the apex of the shoots (in the zone of the whorl). Their needles look like separate clusters (Fig. 14).

There is no unanimous opinion about the mechanism of acute foliage damage by high pollutant levels. We share the opinion that such damage by acid gases is promoted by the photodynamic effect of chlorophyll. The photodynamic (photo-oxidative) effect of chlorophyll that leads to photodestruction of cell components under specific conditions can be concluded from work by Timiryazev (1957). Noack (1920, 1925) was the first to demonstrate the photo-oxidative effect of chlorophyll in plants exposed to acid gases and light. Under the toxic effect of tobacco smoke or sulphur dioxide he observed slight damage to plants in the dark and severe irreversible damage in the light in all cases. In addition, he carried out experiments with white and green halves of *Hemerocallis* leaves. In the dark both halves of these leaves failed to be injured by sulphur dioxide while in the light only the green halves were affected. Noack came to the conclusion that when photosynthesis was suppressed the photochemical energy of

Fig. 14. Regeneration of needles in 2-year-old pine shoots by development of short auxiblasts from dormant buds

chlorophyll was directed towards photo-oxidation of the protoplasm and pigments of chloroplasts themselves.

Noack's results were corroborated by Wehner (1928) who demonstrated an increase in foliage damage following treatment of plants with sulphurous, hydrochloric and nitric acids as well as with ammonia and alkali both in direct light and scattered light. Following this works Krasinsky (1940, 1950) suggested that gas injuries of plants could be classified in terms of the photodynamic effect of chlorophyll as the main factor of acute damage ("burn") followed by acidification of the cellular medium. These data are in agreement with those of Nikolaevsky (1979) who showed that acid gas molecules possess properties of free radicals and can contribute to enhancement of the chain of free radical oxidation processes in plant cells and lead to formation of organic peroxides.

Summarizing the available evidence the mechanism of acute damage can be presented as follows. An acid gas entering the cell via stomata or the cuticle is accumulated in considerable amounts in chloroplasts. High concentrations of the pollutant promote rapid inactivation of enzymes (via pH changes or metal binding) causing an inhibition of the dark phase of photosynthesis, although energy binding by pigments is not inhibited. The energy absorbed by chlorophyll contributes to the photodynamic oxidative processes which form an autocatalytic chain of radical reactions. Increasing photo-oxidation of substances causes destruction of the main protoplast components resulting in cell death. The pigments that provide energy for these photodynamic oxidative reactions are also destroyed in turn because they can no longer bind to the protein–lipid complex. The initial chlorosis, resulting from chlorophyll destruction during the progression of oxidative reactions, is followed by dehydration and destruction of cell components and finally by necrosis.

According to this point of view, the acute damage occurs in the light. Intense visible light has been shown experimentally to cause possible damage to chlorophyll-containing plant cells. The cell damage is enhanced in the presence of any agent, such as chemicals or high or low temperatures, that inhibits photosynthesis. Moreover, cells with suppressed photosynthesis are shown to experience a risk of damage not only from intensive but also from fairly weak light because they lack functional defences against excesses of light energy (Lomagin 1969). The dominant role of light in inducing acute damage has been demonstrated by many workers. Thus, a direct relationship was shown experimentally between the increase in light intensity and increase of foliage damage by sulphur dioxide (Guderian 1979). Plants exposed to hydrogen fluoride and sulphur dioxide were reported to suffer less in the morning and evening and more in the afternoon at a high intensity of illumination (Zimmerman and Hitchcock 1956; Kulagin 1974). Also, increased forest decline has been observed in years of intensive solar radiation (Thiel 1985). Our evidence on performance of plants fumigated by hydrogen fluoride of similar concentrations but under different conditions of illumination also suggests greater plant damage in the light. Even comparatively gas-resistant poplars and willows were shown to suffer damage under overall solar illumination, while such fluorine non-resistant plants as pear and barberry were less injured in shaded chambers.

Many researchers relate light-stimulated damage of plants to stomatal opening which is responsive to light. Exposure to light promotes gas-exchange intensity and hence, the rate of pollutant uptake and its lethal accumulation in cells. Nevertheless, the role of sunlight should not be restricted only to its effect on the photoactive responses of stomata. As reported above this effect is much wider. Furthermore, it is generally accepted that pollutant admission is not only through stomata but also by absorption by ground tissues. The amount of pollutant entering the plant in different ways may vary. Fluorine, chlorine and sulphur dioxide are readily absorbed in the dark with the stomata closed (Benedict et al. 1965). Moreover, stomata may likewise be closed in the light; for instance, under conditions of water deficiency (Kramer and Kozlovsky 1983).

Acute injury by hydrogen fluoride often involves, along with damage to needles, damage to young green shoots of larch and spruce (Fig. 15a,b). The damaged shoots wilt rapidly and dry up. They lose their needles but are kept on the trees for 2 or more years. Young shoots of pine seldom suffer acute damage; their resistance seems to be perfectly provided by leaf lamellae. In the year after spruce damage dormant buds give rise to substituting shoots in the upper part of the 2-year-old shoot and in the zone two to three whorls below the apex of the shoot (the initial phase of their formation can be

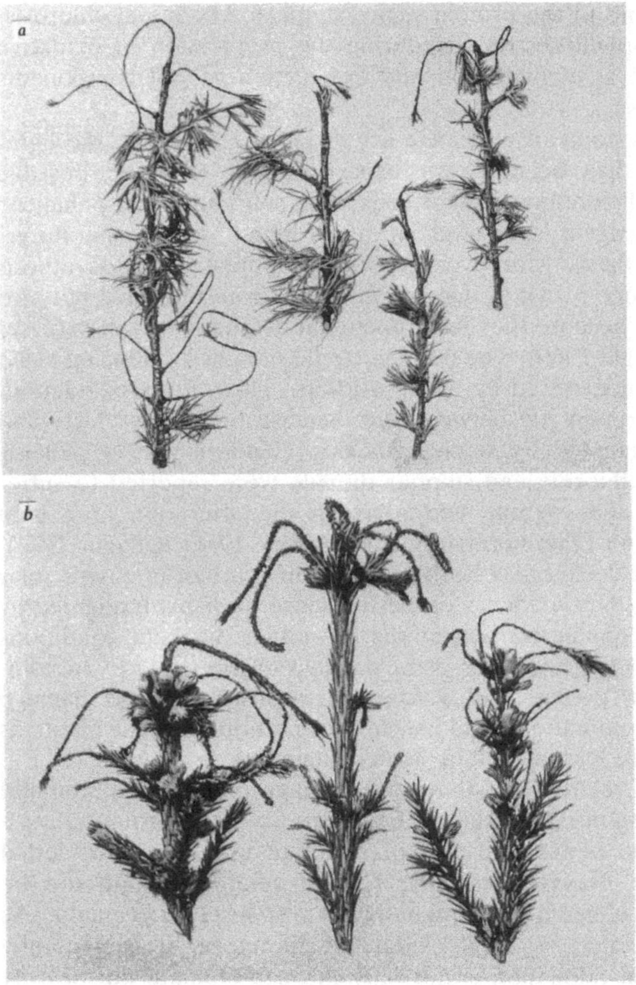

Fig. 15. Young shoots of **a** larch and **b** spruce dried because of the acute damage by hydrogen fluoride in the zone of high fluorine concentration 1 year beforehand

seen in Fig. 15b). Larch regenerates its shoots mainly by transforming brachyblasts into auxiblasts.

The dieback of 1-year-old shoots in winter and in early spring should be also classified as acute damage. It is undoubtedly favoured by frosts and a drop in temperature. In Siberia winter damage of 1-year-old shoots by hydrogen fluoride is typical of all conifers; this is followed by drying up of these irreversibly damaged shoots in May to early June (Fig. 16).

Acid gas injury of plants is stimulated by enhanced air moisture, rainfalls, fogs and dews. Acid rains during the hundred years since the develop-

Fig. 16. One-year-old shoots of spruce (*above*) and larch (*below*) dried because of acute damage by hydrogen fluoride in winter (the pictures were taken in the following June)

ment of this notion have received a great deal of attention in a variety of publications from fundamental research to fiction. They undoubtedly deserve this attention since they constitute a dangerous industrial factor exhibiting injurious effects on the environment. Nevertheless, acid rains do not play as significant a role in mechanisms of acute damage to woody plants by pollutants as enhanced air moisture and fogs do. The degree of plant damage by acid gases is found as a rule to increase with increasing air moisture. It is associated with complete stomatal opening and an increase in the rate of gas absorption in the presence of moisture on the leaf (needle) surface. Thus, the rate of sulphur dioxide precipitation on the wet needle surface, as compared to the dry surface, may rise 10-fold (Smith 1985). Hydrogen fluoride is superior to all other acid gases and photo-oxidants in its water solubility; this stimulates its highly aggressive behaviour under conditions of increased moisture. Keller (1975a) has followed the increase in hydrogen fluoride absorption and its harmful effect on plants under conditions of increased air moisture.

The formation of dew facilitates precipitation of gaseous and solid pollutants on plants. Of special importance are dew droplets in originating corrosive burn in leaves and needles. The drop-like burn of needles can be induced both by high and "non-burning" concentrations of hydrogen fluoride in the atmosphere. It is similar to a solar-heat burn formed by dew droplets but is enhanced many-fold by the effect of fluorine. One idea on a possible gradual increase of pollutant concentration owing to multiple evaporation of rain and dew droplets at leaf edges was suggested 10 years ago (Guderian 1979) but it seems to have only been recalled once since then (Teuchert and Teuchert 1983). It is known at present that rain does not increase fluorine concentration and can even decrease it, both on the leaf surface and in tissues. Contrary to this, dew that has fallen during several consecutive days in clear weather and weak wind promotes the rise in fluorine concentration and appearance of needle necrosis induced by burn.

We have investigated the formation and localization of dew droplets in needles and the appearance of burns. Examining the location of morning dewdrops on needles or, for instance, on grasses, it is easy to notice that they predominate at leaf apices. This is the late phase of drop transformation and is completed by their evaporation. During the first phase the smallest dewdrops evenly cover the whole surface of the leaf; droplets gradually expand and can be seen easily by the naked eye. Leaf burns are mainly linked to larger drops that last longer than others and usually evaporate by the sun's rays.

The formation of large drops only at particular sites on the leaf is determined by its shape, surface structure and intermolecular relationships between contacting media. Dew evaporation (at first minute and small drops) is a reversible process for a certain period. The adjacent layer of air is saturated with water vapour and microcurrents are formed which correspond to the general form and structure of the leaf. A number of vapour particles collide with air molecules and return to the liquid phase. During the next condensation of water vapour, initial dew droplets situated on convex leaves serve as centres of condensation. In broadleaf plants dewdrops collect in the depressions between leaf veins or at each side of the central vein, in the pointed apex, in a dip created by the curved leaf plane and at the leaf edge when the surface is slightly wet. In plants with lanceolate, narrow, linear or needle-shaped leaves large dewdrops are formed in the apex and at points of contact with another leaf or stem and smaller drops are formed in the upper part of arched curvatures. Drops are also formed at sites where cuticle integrity has been disturbed (points of

Fig. 17. Drops of dew on the tips of pine needles. On the *right* is a dew droplet below the apical damage of the cereal leaf.

wetting). In the stiff, single needles of pine most large dewdrops are formed in the apical region (Fig. 17) and it is here that initial burns mainly appear.

Burns induced by a pollutant at sites of large dewdrop formation result from the relative stability of the droplet. Following the repeated appearance and evaporation of droplets containing even a little of the toxic compound, its concentration may increase in a certain place to a level sufficient to produce a burn. It is quite possible that the corrosive burn is combined with the effect of direct sun rays and with varying temperatures that result from the formation and evaporation of dew. If there is already an apical burn the dewdrop is formed immediately below it (see Fig. 17) and the injured spot expands. In the case of a severe local burn in the middle of a needle with injured epidermis and mesophyll the isolated part dies and the basal part, unlike in acute or cumulative toxic effects, remains alive for a long time. A dark strip, sometimes also with a light line, appears at the boundary between the burn and the living part of the needle. These belts are particularly distinct in pine needles (including those that are dead) but they are absent or seldom seen in spruce. The drop-like burn is of less importance to woody plants than are acute and cumulative damages. However, it still poses a considerable threat to species with needles that are many years old, especially when trees are on the verge of irreversible weakening.

Cumulative damage of needles is induced by a long-term effect of fairly low concentrations of acid gases and forms a part of the general process of slow degradation in plants. In contrast to acute damage the pollutant dose needed to inhibit photosynthesis is slow to be accumulated and therefore the photo-oxidative effect of chlorophyll is of secondary importance. The gradual entry of small amounts of fluorine allows the plants to detoxify themselves and to translocate the fluorine along transpiration and phloem pathways. In acutely damaged trees the degree and rate of destructive processes are largely determined by the resistance of the photosynthetic apparatus and the regulatory systems of the cell; of greater importance under the long-term effect of low concentrations are defence-adapting responses that contribute to an increase in organism resistance to stress factors and that render the pollutant harmless. A gradual decrease in adaptive ability leads to a steady increase in metabolic disorders. The visual signs of cumulative damage to needles are expressed in terms of growth suppression, abundant necrosis, reduction of needle life in shoots of previous years and dieback of separate needles in 1-year-old shoots.

A single needle suffering from cumulative fluorine damage gradually declines and necrosis, as in the case of the burn, spreads from the apex to the base. However, in contrast to the burn, necrosis extends continuously and the dark belt separating the living part of the needle from the dead is hardly visible. The appearance of such apical necrosis can be accounted for by the fact that low levels of hydrogen fluoride entering the needle are translocated towards the apex in the transpiration stream. The pollutant gradually accumulates here disrupting metabolic processes and leading to cell death. According to Ilkun and Bukolova (1974) fluorine causes obstruction of vessels and mesophyll destruction, thus preventing entry of water and nutrients to the needle apex and resulting in needle death. The cumulative damage by fluorine induces gradual drying of branches and of the whole tree. It is much more widespread than acute damage, can be observed at a distance of several tens of kilometers from the source of pollution and covers vast areas.

No special investigations were carried out on latent damage of conifers, although we encountered this type of damage when assessing maximum permissible fluorine concentrations in tree needles. A comparison between experimentally determined permissible fluorine content in needles and that actually found in trees growing outside the zone of slight effect of fluorine emission revealed identical accumulation of fluorine. It is interesting that while no visual and physiological changes occur in needles under the permissible fluorine level, even a slight increase in this level produces detectable physiological changes (in most cases statistically significant) in trees. These trees may be suggested to pass into a state of "unsteady balance", induced by the effect of latent fluorine impact.

A 10-day treatment of young pine, larch and spruce trees with a very low hydrogen fluoride concentration (2.3 $\mu g/m^3$ supplied for 2–3 h/day) revealed an interesting phenomenon. In the year of fumigation there were no visible signs of needle injury; however, in the next growing season hypertrophic needles appeared on young shoots of pine and spruce and in the whole crown of larch. These needles lacked any symptoms of chlorosis and necrosis. In subsequent years the trees had normal-sized needles. Fluorine appeared to exhibit a stimulating effect on these plants. Similar effects of fluorine have been reported by other authors (Guderian 1979). Available evidence also suggests an increase in productivity of some crops, e.g. oat, carrot, bean, hemp, cabbage, potato, poppy and pea, when treated with small doses of fluorine (Berenshtein 1953). Further investigations are needed to elucidate the nature and mechanism of fluorine-promoted plant growth.

3.2 Pathways of Fluorine Entry into the Plant; Its Translocation, Localization and Removal

Pollutants enter the plant mainly through stomata during the process of gas exchange. In addition, gaseous pollutants can penetrate through leaf epidermal cells and softwood shoot cells and through the bark of the trunk and branches. Moreover, fluorides can enter the plant through the root system. Plants start suffering damage by injurious gases from the moment of gas penetration through the stomatal pore and the cuticle. However, pollutants must first be absorbed by the foliage surface. The process of absorption of polar gases (HCl, SO_2 and HF) has been considered at length by Ilkun et al. (1983) who propose the following mechanism for gas molecule absorption by leaves. Absorption sites on the leaf surface are usually filled with water molecules but when polar gas molecules appear in the air they start to be absorbed along with the water molecules at these sites. The process continues until there is a dynamic balance between the concentration of gases in air and their content on the leaf surface (in a bound state). The microtopography of the leaf surface is of great importance for the process of gas binding as a highly partitioned epidermis increases the leaf surface interaction with the ambient air and correspondingly allows an increased amount of gases to be absorbed.

Absorbed chemical pollutants are dissolved in the layer of water on the leaf surface and the ions formed enter epidermial cells through the cuticle. They move via the free space into neighbouring mesophyll cells and conducting tissues, along which they are transported into other organs and tissues (Ilkun 1971a,b; Negrutsky et al. 1981). The accumulation of pollutant ions in cells reduces the difference between the pollutant concentration in the atmosphere and in the leaf; this appears to reduce the rate of further pollutant uptake. If the pollutant continues to enter cells it means that the difference in concentration must be kept at the same level. And for this to occur pollutant ions entering the cells must be constantly bound or removed (Negrutsky et al. 1981).

Fluorides administered to plants through the root system, as reported in various studies (Ilkun 1971a; Walberg 1976; Cicák 1979), result in the same damage as in the case of exposure to atmospheric fluorine. The pattern of visible damage also fails to change when fluorine enters the plant organism through bark of the trunk, as will be discussed later. Therefore, fluorine ions entering the plant by any one of a number of pathways induce similar damage to plants.

When a pollutant enters a cell the cell's defence response is primarily aimed at detoxification or isolation of the pollutant (Durmishidze et al. 1981). Detoxification usually implies chemical transformation of the pollutant by metabolic processes leading to a reduction in pollutant toxicity. The most common mechanisms of detoxification are: neutralization by synthesis of oppositely charged ions, involvement in reactions with enzymes and formation of conjugates (Ilkun 1971b; Durmishidze et al. 1981). Which of these methods of detoxification is characteristic for fluorine is not yet known with certainty. There is an opinion that the main way of rendering fluorine ions harmless is for them to be bound with free cations (Ilkun and Motruk 1980). Fluorine involvement in metabolic processes has not been shown for certain and can only be assumed from the data on increased productivity in some plants when given small doses of fluorides.

Another way to reduce the injurious effect of a pollutant is to remove it from the plant organism and release it into the environment. Fluorides have been shown to be removed by desorption by leaves, leaching by precipitation, formation of root exudates, annual defoliation and water transpiration (Ilkun 1978; Ilkun and Motruk 1980; Ilkun et al. 1983). Furthermore, woody plants are able to diminish the activity of absorbed pollutants by storing them in dead or poorly functioning tissues of the trunk and roots. Our experiments with hydrogen fluoride showed a high mobility of fluorine in the conifer, considerable flow of fluorine from metabolically active tissues (needles, phloem), fluorine accumulation in the outer bark during the recurring process of detachment of the peripheral layers of active phloem as periderm and accumulation in sapwood and heartwood.

The problem of whether fluorine is translocated in plants has been solved quite readily. Yet in the 1960s possible transport of fluorine from leaves to stems and roots and its redistribution between different-aged leaves was overlooked (Hitchcock et al. 1963; Benedict et al. 1964), though 10 years later transport within plant organisms, both deciduous and coniferous trees, had been considered (Keller 1974a,b, 1975b; Kronberger et al. 1978). In our opinion the translocation of fluorine from metabolically active tissues should be considered as one of the main methods of detoxification.

We have followed the different ways fluorine enters the plant: through needles, the root system and trunk. The aim of the research was first to see how each of these pathways contributes to the total process of pollutant accumulation and second, to follow fluorine translocation within the plant from each of the pathways of entry. Special attention was paid to individual organs involved in the process of fluorine release into the environment and to the role of different tissues in storing fluorine.

Fluorine accumulation in tissues and tree decline following simultaneous entry of fluorine through needles, shoots, trunk, and also partially via roots, was determined in fumigation chamber experiments. The fumigation of 10- to 12-year-old trees with 0.2–0.3 mg/m^3 hydrogen fluorine at a rate of 20 h/day for the whole of the growth period resulted in heavy needle necrosis and accumulation of a high level of fluorine by the end of the growing season. Higher levels of fluorine were also found in the phloem and sapwood of the trunk and branches and in the roots as compared to control trees (Fig. 18). Fluorine accumulation in roots and sapwood results, to a certain degree, from fluorine redistribution, i.e. downward flow via the phloem from needles and upward flow via sapwood from the roots. The experimental results obtained, compared to the data on fluorine accumulation in tissues of young trees growing close to the source of fluorine emission (see Fig. 12), indicate that the needles accumulate the highest amount of fluorine while other organs are less loaded with fluorine.

This next experiment was the most interesting with respect to elucidating the pathway and rate of downward flow of fluorine from needles and its localization in tree organs and tissues. Film chambers of 20 dm^3 were attached to the top of pine trees (20- to 25-year-old polewood) (see Fig. 2) and were supplied with 0.1 mg/m^3 hydrogen fluoride. The chambers were removed from the trees after 3, 15, 30, 40 and 75 days and the fluorine content was assayed, mainly in needles, phloem, outer bark, sapwood and roots (Table 9).

The elevated fluorine content in needles outside the chamber during the first 3 days, compared to the concentration in phloem, can be accounted for by leakage through ventilation holes. As already mentioned the needles absorb hydrogen fluoride most actively during gas exchange. However,

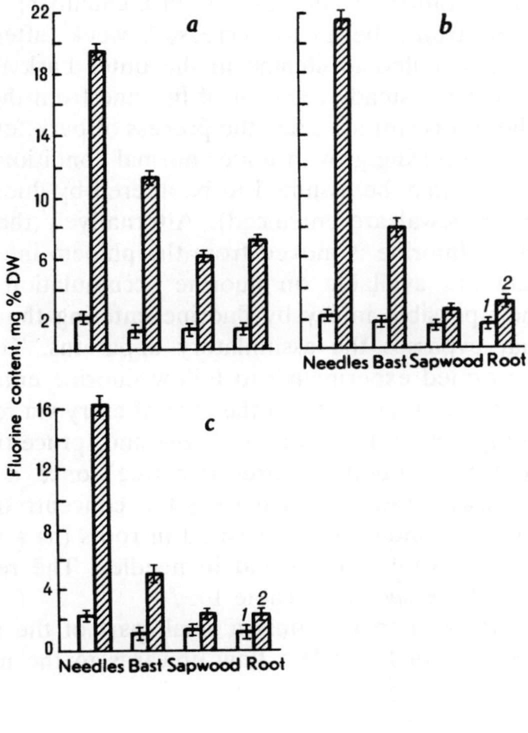

Fig. 18. Fluorine accumulation in 12- to 15-year-old trees of **a** pine, **b** larch and **c** spruce following fumigation in chambers. *1* Control; *2* experimental value

Table 9. The content of fluorine (mg% DW) in organs and tissues of 20- to 25-year-old pine trees following local fumigation of the crown top for various periods of time with 0.1 mg/m^3 hydrogen fluoride

Organ/tissue[a]	Length of fumigation (days)					Control
	3	15	30	40	75	
Needles						
in chamber	7.06 ± 0.05	13.19 ± 0.07	18.48 ± 0.20	21.89 ± 0.13	22.11 ± 0.07	1.84 ± 0.03
outside chamber	3.04 ± 0.04	3.40 ± 0.05	3.50 ± 0.02	4.13 ± 0.05	5.80 ± 0.04	1.84 ± 0.03
Phloem	2.20 ± 0.08	4.97 ± 0.14	9.41 ± 0.06	–	9.03 ± 0.05	1.22 ± 0.02
Outer bark	1.05 ± 0.13	1.83 ± 0.11	2.42 ± 0.05	3.23 ± 0.06	5.55 ± 0.08	0.96 ± 0.02
Sapwood	1.29 ± 0.12	2.16 ± 0.05	3.15 ± 0.09	–	6.48 ± 0.06	1.12 ± 0.03
Roots	1.37 ± 0.14	2.30 ± 0.03	6.43 ± 0.07	–	7.56 ± 0.06	1.17 ± 0.07

[a] The phloem, outer bark and sapwood were analysed for the height of the trunk and roots as a whole; see also Table 12.

subsequently the content of fluorine increases to a greater extent in the phloem and then in the sapwood, although in sapwood the elevation of fluorine content is much inferior to that in phloem since it occurs mainly by the upward flow of fluorine from the roots. This process was followed in a particular set of experiments.

Of special interest is the increase in fluorine content in the outer bark; this is a slower process than in other tissues. The rate of increase of fluorine concentration in needles (within a chamber) and in the phloem, along the whole trunk, began to decrease 2 weeks after the start of the experiment while it failed to change in the outer bark throughout the 75 days. This indicates a steady removal of fluorine from the phloem into the outer bark where it is stored. Since the process of bark formation cannot be continuous in pine during growth under normal conditions the rate and rhythm of this process may be assumed to be altered by fluorine (when phloem formation and renewal are enhanced). Alternatively there may exist another way in which fluorine is moved from the phloem into the outer bark. Actually no data are available on fluorine accumulation in woody plant tissues and their possible injury by fluorine entering the plant organism by a method that bypasses the assimilatory apparatus. To further investigate this we performed experiments to follow fluorine entry into the plant via the root system and the bark. In the case of entry via roots the soil within the foliage canopy of 10- to 13-year-old pine and spruce trees was watered with sodium fluoride solution (2 litres per tree) once a week for 3 months (June–August), gradually increasing the concentration from 0.01 to 0.1%. The fluorine content was measured in roots (as a whole), phloem and sapwood (of the whole trunk) and in needles. The results expressed as mg% dry weight are shown in Table 10.

In both species only a small part of the fluorine entering the roots is moved with the xylem flow upwards to the needles. Following a 3-month

Table 10. Fluorine accumulation in 10–13-year-old trees watered weekly with 2 l of sodium fluoride in concentrations increasing from 0.01 to 0.1% during June–August

Species	Object of study	Content of fluorine (mg% DW)	
		Control	Experiment[a]
Pine	Needles	2.00 ± 0.097	4.17 ± 0.126
	Phloem	1.40 ± 0.058	3.65 ± 0.141
	Sapwood	1.80 ± 0.068	4.53 ± 0.128
	Roots	1.77 ± 0.088	16.95 ± 0.092
Spruce	Needles	2.17 ± 0.071	4.08 ± 0.101
	Phloem	1.97 ± 0.095	4.97 ± 0.088
	Sapwood	1.07 ± 0.067	2.45 ± 0.109
	Roots	1.88 ± 0.60	20.28 ± 0.020

[a] A number of treated trees were used for further observations and experiments.

toxication period fluorine content was 12–27%, 20–25% and 21–25% of the root concentration in the sapwood, needles and phloem respectively. It takes the plant much longer to accumulate fluorine in the needles to a concentration inducing their death by this method than by fumigation of the crown with 0.01 mg/m^3 hydrogen fluorine. We could not find any visible symptoms of injury in the treated trees in the year of the experiment and only in the following year during the early growing period did one-half of the trees exhibit signs of chlorosis and necrosis (about 10% damage). During the next year treatment of the roots with sodium fluoride solution of the same gradually increasing concentration (during the same 3 months) produced severe damage of needles and even irreversible weakening of some trees, which later declined. The fluorine concentration in the soil beneath the treated trees was 0.118% of dry weight by the end of the growing season (0.025% in control).

It is significant that when fluorine enters via the root system the speed of its transport by downward flow also remains rather high; this is evident from the ratio of fluorine content in needles:phloem in pine and spruce (4.17:3.65 and 4.08:4.97). It should be noted that the fluorine content in the phloem does not exceed 9.5 mg% dry weight even if it is very high in the needles (18–22 mg% dry weight) (see Table 9). The higher content can only be registered when the pollutant enters the phloem through the thin outer bark.

To examine the possibility of and degree of hydrogen fluoride penetration through the bark of trees of different age (young, polewood and mature) and subsequent fluorine transport, contact chambers with a high hydrogen fluoride concentration (0.8 mg/m^3) were attached to pine trunks (see Fig. 2). In 8–10 days the needles of young trees became yellow above the chambers and after 12 days the damage embraced the whole crown. The

Table 11. Fluorine accmulation in the sapwood, phloem and roots of 12–15-year-old pine trees during local fumigation of the middle part of the trunk with hydrogen fluoride (0.8 mg/m^3) for 12 days

Object of study	Sample	Fluorine content (mg% DW)
Phloem		
Experiment	Above chamber	6.35 ± 0.096
	Within chamber	21.33 ± 0.174
	Below chamber	29.65 ± 0.217
Control		2.35 ± 0.067
Sapwood		
Experiment	Above chamber	5.28 ± 0.117
	Within chamber	10.25 ± 0.200
Control		1.38 ± 0.065
Roots		
Experiment		1.55 ± 0.106
Control		1.08 ± 0.060

fumigation chambers were kept attached to the trunks of polewood trees for 20 days at the beginning of the growing season. No changes were registered in the needles both above and below the chamber during this period; the damage appeared in the second part of the growing period and was quite typical of that caused by fluorine. The chambers were kept on mature pine trunks during 3 months in the summer and only during the next spring could one see necrotic tips to needles throughout the whole crown. The experiments demonstrated the reduction in fluorine penetration through the outer bark with tree age.

The larger part of fluorine entering the trunk is localized in the phloem within the chamber and below it, having moved there along the phloem flow. Fluorine appears to penetrate into the living tissues within the chamber by diffusion from the oversaturated phloem. The content of fluorine in the roots is only 43% above that of the control; this could be explained by the short duration of the experiment (12 days) (Table 11). It is noteworthy that the level of fluorine in the sapwood above the chamber was twice that within the chamber, while its concentration in the downward phloem flow was 40% higher than in the zone of direct entry. It is reasonable to assume the existence or appearance of defence mechanisms in trees to restrict the penetration of pollutants by upward xylem flow into the crown.

Similar chambers were attached to trees and used to study the pathway and rates of fluorine removal from and storage in needle tissues. The chambers were hung on branches in the upper part of the crown of 20- to 25-year-old pine trees and used in fumigation experiments for 15 days before being removed. The content of fluorine in the needles, phloem, sapwood, outer bark and roots was measured immediately after the removal of the

Table 12. The content of fluorine (mg% DW) in organs and tissues of 20- to 25-year-old pine trees at different times after fumigation

Time after fumigation (days)	Needles		Phloem	Outer bark	Sapwood	Roots
	In chamber	Outside chamber				
0	16.35 ± 0.08	3.44 ± 0.03	5.57 ± 0.06	1.75 ± 0.06	2.40 ± 0.04	2.85 ± 0.16
3	16.04 ± 0.07	3.12 ± 0.04	4.65 ± 0.05	1.92 ± 0.11	2.29 ± 0.10	2.44 ± 0.08
10	13.02 ± 0.08	2.33 ± 0.05	4.46 ± 0.11	2.84 ± 0.07	2.42 ± 0.07	3.15 ± 0.12
30	12.42 ± 0.34	2.28 ± 0.07	4.32 ± 0.09	3.05 ± 0.07	2.15 ± 0.08	3.45 ± 0.07
60	10.60 ± 0.27	2.22 ± 0.04	2.77 ± 0.08	3.61 ± 0.07	2.21 ± 0.14	4.55 ± 0.08
330	7.34 ± 0.13	1.91 ± 0.04	1.48 ± 0.09	5.39 ± 0.08	2.52 ± 0.11	2.44 ± 0.06
Control	1.84 ± 0.03	1.84 ± 0.03	1.22 ± 0.02	0.96 ± 0.02	1.12 ± 0.03	1.17 ± 0.07

chambers, at 3, 10, 30 and 60 days later and at the beginning of the growing season (Table 12).

The results allow the following suggestions to be made about the relative resistance to fluorine of coniferous trees. First, fluorine is more rapidly eliminated from the needles and phloem which, in turn, are the first to become saturated during fumigation. A high residual content of fluorine in the needles inside the chamber can be accounted for by their severe damage and partial dieback when fluorine reaches an extremely high concentration. Second, fluorine removal by transpiration into the environment from needles is an important, though not major, part of the mechanism for diminishing the fluorine concentration in needles. The content of hydrogen fluoride in the atmosphere of the stomatal cavity and mesophyll intercellular space is constantly maintained, by diffusion, at the level of its concentration in the ambient air. Fluorine is dissolved in water evaporating from the surface of mesophyll cells and diffuses in this form through stomatal pores but a certain amount of it penetrates into these cells. An excess accumulation of fluorine in the mesophyll is prevented by its flow, together with metabolites, into the phloem. The priority of decreasing fluorine concentration in needles by phloem transport can be suggested by the fact that the reduction in total amount of fluorine in needles (outside the chamber) and in phloem during the post-fumigation period correlates with the increased total fluorine content in the outer bark and roots.

Third, a uniform and significant rise in fluorine concentration in the roots during the post-fumigation period and its later decline has little or no effect on the dynamics of fluorine appearance in the sapwood. It suggests, on the one hand, a restrained flow of fluorine from the roots into the crown (xylem flow) and, on the other hand, rapid fluorine removal by the roots into the soil. The process of fluorine removal by roots is likely to occur at the end of the growing season or in spring, though the latter is less probable. Fourth, the content of fluorine found in the sapwood immediately after the 15-day fumigation then remained almost unchanged for 1 year (change from

2.21 to 2.52 mg% dry weight), while it successively increased or decreased in needles, phloem, outer bark and roots. This implies that the residual, or stored, amount of fluorine in the pine sapwood is in the range of 2.0–2.2 mg% dry weight. Analysis of trees in the zone of high fluorine concentration in the vicinity of the aluminium smelter showed the residual amount of fluorine in larch and spruce sapwood to be twice as high. Fluorine is also accumulated in heartwood. Under many years' exposure to fluorine the content in heartwood should not be lower than the residual amount in sapwood, at the expense of which this wood is produced. However, there are no credible data available on fluorine accumulation in heartwood and on the dynamics of this process.

The outer bark is the main site of accumulation and deactivation of fluorine, as has already been mentioned. Fluorine enters this dead part of the bark until its concentration is above the normal level found in phloem. In the experiment the content of fluorine in the outer bark increased more than threefold and was 5.5 times greater than the control level at 330 days after cessation of fumigation. Fluorine enters the outer bark only from the phloem since the hydrogen fluoride concentration is similar to the atmospheric level during local fumigation of branches.

In forests exposed to aluminium plant emissions the outer bark of conifers absorbs a substantial amount of fluorine directly from the atmosphere. Fluorine absorbed from the atmosphere predominates in the thin bark of branches and at the top of the trunk; the proportion of fluorine accumulated from the air decreases with thickening of the outer bark and fluorine enters the outer bark mainly from the phloem.

To establish the maximal accumulation of fluorine in outer bark and its localization across bark thickness outer bark was analysed, layer by layer, in 40- to 60-year-old pine and larch trees growing in the zone of high fluorine concentration (0.05–0.15 mg/m^3) and following 15 years of aluminium smelter operation. It was found that the inner part of the outer bark may contain almost twice as much fluorine as the phloem (see Table 13).

Concentration of fluorine in the external part of the outer bark is higher still and results from severe contamination of the bark surface with dusty particles containing fluorides.

Fluorine is evenly distributed throughout the inner part of the bark. Its concentration appears to be similarly higher in larch and pine stands polluted with fluorine-containing emissions. Under these conditions the maximum concentration of fluorine in the inner part of the bark that is isolated from penetration by aerosol fluorides is 7–8 mg% dry weight. During the experimental conditions of local fumigation of the part of the trunk with a high concentration of hydrogen fluoride the bark can show higher fluorine saturation.

When fluorine enters the plant organism only through the needles or root system, under experimental conditions, its accumulation in various tissues varies greatly. These differences partly diminish with time. In areas

Table 13. Fluorine accumulation in the phloem and in the external and internal parts of the outer bark in pine and larch in the zone of high concentration of fluorine (0.05–0.15 mg/m^3)

Species	Object of study	Fluorine content (mg% DW)	
		Control	Zone of high concentration
Pine	Phloem	1.50 ± 0.58	4.32 ± 0.16
	Outer bark		
	External part	0.98 ± 0.02	18.81 ± 0.31
	Internal part	0.76 ± 0.02	7.63 ± 0.10
Larch	Phloem	1.25 ± 0.08	4.10 ± 0.15
	Outer bark		
	External part	0.92 ± 0.03	11.91 ± 0.26
	Internal part	0.90 ± 0.03	7.42 ± 0.12

constantly polluted by aluminium smelter emissions fluorine, as an aggressive pollutant, penetrates into the plant by all available pathways: through stomata, the cuticles of leaves and of young shoots, the outer bark of the trunk, epiblema and secondary bark of the root. The content of fluorine in living tissues is kept constant in this case. The largest fluorine concentration is found in the needles with a lower concentration in the phloem and the least in the sapwood and roots. The content of fluorine in root tissues of young and slightly weakened premature and mature trees is usually somewhat higher than in the sapwood; its content in the sapwood increases appreciably with physiological weakening. If an average fluorine concentration in the living tissues of a 20- to 40-year-old pine is taken as a unit, fluorine content would be 1.8–2.0 in needles, 0.9–1.0 in phloem, 0.5–0.7 in root xylem and phloem and 0.5–0.6 in sapwood.

Summarizing the data presented, we may conclude that fluorine is far from passively accumulated in the damaged tissues of the plant. The process is much more complex and involves fluorine translocation and redistribution within the plant, flow from metabolically active tissues, neutralization and possible storage inside the plant and removal into the environment (Fig. 19). Approximate relationships between these processes can be illustrated by the thickness of the dark arrows in Fig. 19. As early as during the first stages of fluorine toxicosis the physiologo-biochemical mechanisms of defence are activated; these prevent, in particular, the movement of fluorine into the crown by upward flow.

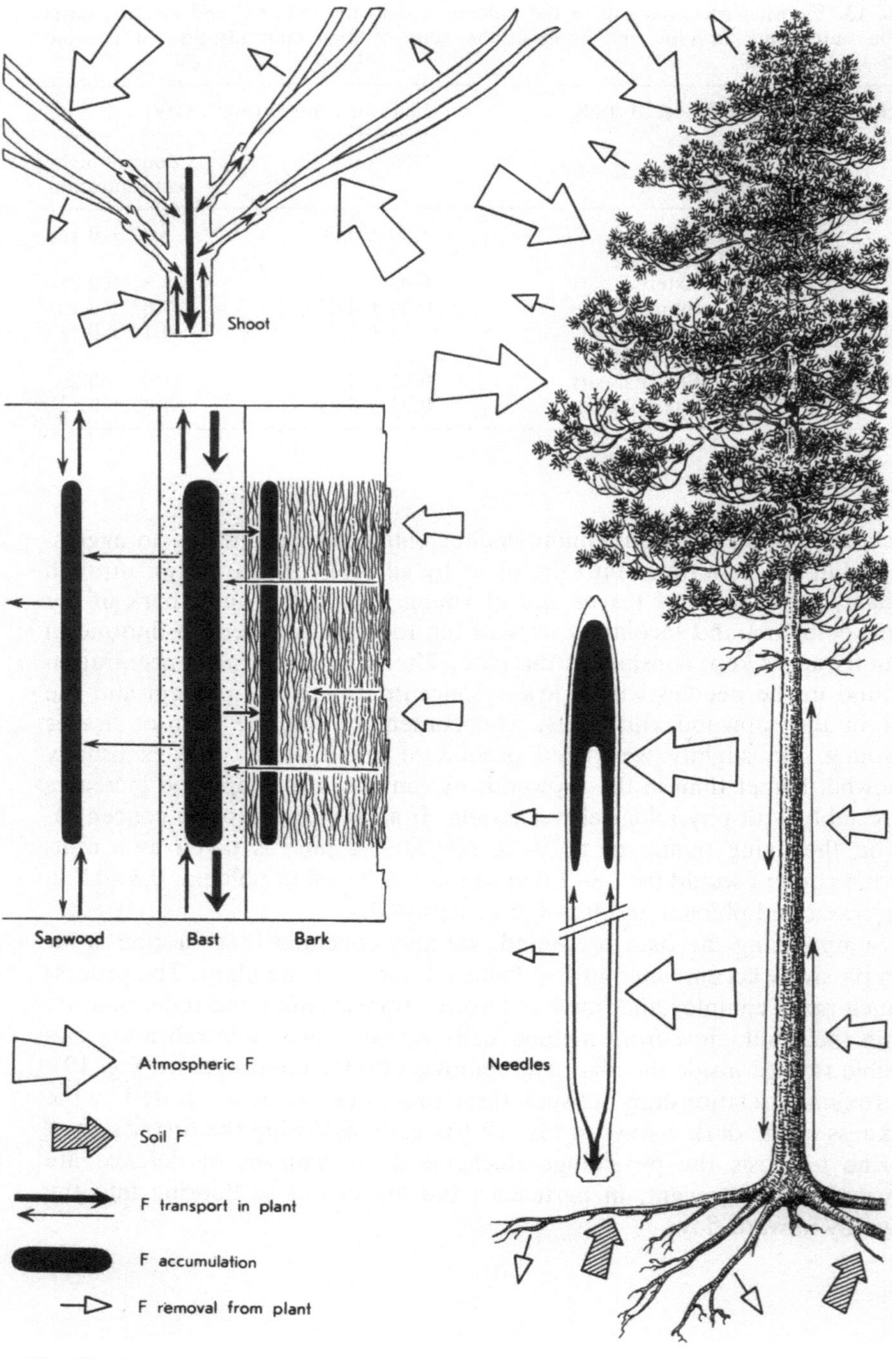

Fig. 19. A scheme of fluorine entry into the plant, its transport and localization in tissues and removal from the plant

3.3 The Effect of Hydrogen Fluoride on Physiological-Biochemical Processes in Conifers

Fluorine and its compounds exert a diverse influence on physiological-biochemical processes of plants. Fluorine anions penetrate readily into the cells and organelles and may have a toxic effect even at low concentrations. It is recognized that the presence of fluorides in the ionic form in cells and the possible high accumulation, primarily in chloroplasts, is responsible for high fluoride toxicity in plants (Chang and Thompson 1956; McNulty and Newman 1956, 1961; Peters and Shorthouse 1964; Jacobson et al. 1966; Torgasheva 1984). Calcium and magnesium binding in the vacuolar membrane (Ling-Ling and Miller 1972), disruption of cellular membranes (Zwiazek and Shay 1988a), changes in DNA and RNA structure (McCune and Weinstein 1971), inhibition of ATP synthesis (Chang 1968) and changes in the activity of a number of enzymes including enolase (Warburg and Christian 1942), acid phosphatase, phosphoglucomutase (Ilkun 1971b; Zwiazek and Shay 1988b) and dehydrogenases (Slater and Bonner 1952) are also initially induced by fluorine.

Most of the research devoted to physiological-biochemical changes induced by gaseous fluorides has been performed with plants damaged to a certain extent. However, it is necessary to have knowledge about the dynamics of physiological processes in plants to elucidate the impact of fluorine on pathological and defence/adaptive responses in plants. We have followed the changes in a variety of physiologo-biochemical features in conifers fumigated with hydrogen fluoride at different stages of damage: at the onset of fumigation when visible signs of injury are still lacking and at slight, moderate and severe extents of needle damage.

Trees, 12- to 15-years-old, of Siberian larch, common pine and Siberian spruce were used in the trials. The trees were exposed to 0.2–0.3 mg/m^3 hydrogen fluoride in fumigation chambers for 18–19 h/day. The total length of fumigation varied because of the irregular appearance of the same damage in different species (Table 14). The period of time within which the initial degree of damage appeared (lacking visible signs) was determined experimentally and was taken as the term of fumigation when there were no visible signs of damage on one day but they appeared on the next day. This period was 2, 3–4 and 5–6 days for larch, pine and spruce respectively. The content of fluorine in needles was 1.6–2.2 times above the control level during this period. Under slight, moderate and severe degrees of damage the amount of necrosis in needles of the crown was 10, 40 and 70% respectively and there was 2.3–4.3 times more fluorine in the remaining green needles than in those of the control.

The fluorine concentration supplied to the chambers failed to induce irreversible weakening of the trees. However, longer exposure under these concentrations resulted in tree decline in the following year. In fumigation

Table 14. Needle damage and fluorine accumulation in needles of 12- to 15-year-old trees

Degree of damage (visual)	Larch		Pine		Spruce	
	Duration of fumigation (days)	F content[a] (mg% DW)	Duration of fumigation (days)	F content (mg% DW)	Duration of fumigation (days)	F content (mg% DW)
No signs of injury	1–2	4.63 ± 0.14 / 1.83 ± 0.03	3–4	3.68 ± 0.16 / 2.08 ± 0.08	5–6	2.95 ± 0.06 / 1.97 ± 0.06
Slight	4–7	5.28 ± 0.15 / 2.05 ± 0.08	8–9	4.90 ± 0.19 / 1.95 ± 0.08	15–16	6.00 ± 0.06 / 1.83 ± 0.07
Moderate	8–13	6.47 ± 0.15 / 2.08 ± 0.10	20–25	6.83 ± 0.21 / 2.17 ± 0.09	35–40	7.93 ± 0.14 / 1.93 ± 0.06
Severe	14–20	6.77 ± 0.13 / 2.32 ± 0.03	40–45	6.85 ± 0.27 / 2.25 ± 0.04	60–70	8.10 ± 0.15 / 1.92 ± 0.08

[a] The numerator indicates the experimental value and the denominator the control value.

chambers the state of irreversible weakening of the tree was characterized as follows. In larch, all auxiblasts died and living brachyblasts (less than 30%) with partially green needles remained only on lower branches close to the trunk. Spruce suffered lethal damage to a larger number of growing shoot buds and to all 1st- and 2nd-year needles. Injured (necrotic) needles from the previous year's growth remained only on the lower part of the crown. The same symptoms were observed in severely weakened pine, though unlike spruce the dead needles remained on branches for longer. Irreversible weakening of trees of all three species follows fluorine accumulation in the needles up to 20 mg% dry weight or more. Such trees failed to recover. Sprouting of a few dormant buds during the next spring could not prolong the life of these trees.

Our experiments were also aimed at following changes in particular steps of tree metabolism during various stages of tree damage prior to the period of irreversible weakening and decline. Physiologo-biochemical traits were determined in needles lacking any visible sign of damage, i.e. necrosis, as in the fluorine accumulation studies and these could be used to characterize the condition of the tree. An average sample from three to five trees was analysed. Trees of the same age placed in the same chambers but without fluorine fumigation served as a control. Physiological-biochemical characteristics were measured by conventional methods: water content was determined by dry and fresh weight variation (Ermakov 1972); peroxidase activity was monitored by the method of Boyarkin (1951); pigment content was analysed by use of a spectrophotometer, Sph-10 (USSR), (Shlyk 1965); respiration intensity and reducing sugars were measured according to the methods of Valter et al. (1957); and the content of ascorbic acid was measured by potassium iodate reduction, total organic acids by alkaline solution titration, total and protein nitrogen by Nessler reagent (Pleshkov 1976), potassium and sodium by flame photometry (Rinkis and Nollendorf 1982) and calcium and magnesium by the atomic absorption method (Prise 1976).

3.3.1 Reduction–Oxidation Processes in Needles Affected by Fluorides

To maintain a balance in reduction–oxidation processes in cells is of primary importance under exposure to acid gases which induce a shift towards acidity in the intercellular medium. Since most hydrolytic enzymes are optimally active under acid conditions, a pH decrease in the cellular medium stimulates their activation; this could result in further oxidation in cells. The shift in metabolism of cells towards oxidative processes can produce a lethal effect even though energy substances are available (Girs 1982).

The activation of peroxidase, induced by unfavourable conditions, is a characteristic response of plant cells which allow the oxidative processes to function at the normal level (Nikolaevsky 1968). It is a common concept that the main function of peroxidase is to defend an organism against the adverse effects of peroxides (Rubin and Ladygina 1974). Moreover, peroxidase is involved in many metabolic transformations in the cells (Sudachkova 1977; Butt 1980). Rachkovskaya and Kim (1980) believe that the defence role of peroxidase, when plants are exposed to acid gases, is associated not only with elimination of the toxic effect of peroxides but also with its functioning as a terminal oxidase. According to Keller (1974b, 1976) peroxidase that is sensitive to atmospheric fluorides can allow detection the negative impact of fluorine on plants well before the appearance of visible damage. Therefore, it has been recommended that peroxidase activity be used to map areas of insidious effect of pollutants. A comparative testing of various methods (turbidity test, buffer capacity of needles, peroxidase activity) for their sensitivity in detection of the toxic effect of acid gases revealed that analysis of peroxidase activity was the most sensitive test.

Tests yielded steady differences in peroxidase activity of conifers under investigation. Prior to the appearance of visible injury the activity of the enzyme increased up to 200% of the control activity in larch, 225% in pine and 135% in spruce. Then it declined to some extent in slightly damaged needles (considerable decline was recorded only in spruce); however, it was still above the control activity. With a further increase in needle damage the enzyme activity either increased or remained at the same level. In the case of severe needle damage it was 500% of the control activity in larch, 200% in pine and 170% in spruce after an initial drop (Fig. 20).

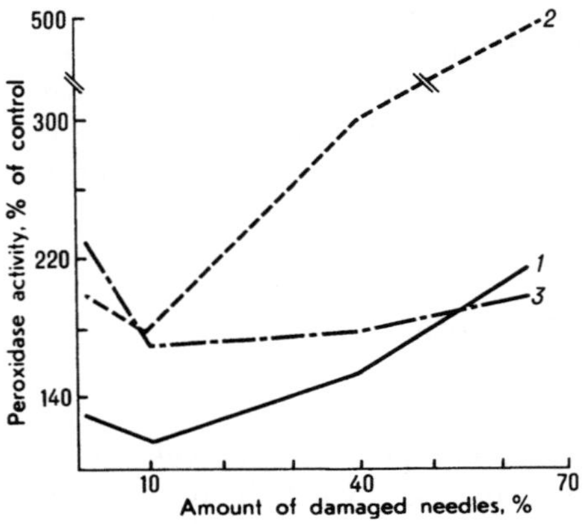

Fig. 20. Changes in peroxidase activity in needles of pine (*1*), larch (*2*) and spruce (*3*) following varying degrees of damage by hydrogen fluoride

Acid gases promote initiation of free chain radical oxidation in the light to produce organic peroxides (Nikolaevsky 1979). The production and accumulation of peroxides are likely to contribute to substrate activation of peroxidase, which is known to use organic peroxides via catalytic action as a source of active oxygen. Ivanova and Rubin (1962) have reported the promotion of oxidative properties of peroxidase as its activity increases. Consequently, acid gases can enhance the functioning of peroxidase as a terminal oxidase. It is likely that, along with inhibition of other oxidases, adaptive reorganization of the oxidative apparatus may occur to prevent disorder to the process of respiration. Nevertheless, excessive activation of peroxidase followed by inhibition of respiration in severely damaged larch produces an adverse effect since it elevates the oxidative potential of the cells.

Entry of fluoride ions into assimilatory organs of plants is responsible for changes in the performance of certain oxidative systems of respiration; fluorides either stimulate or inhibit respiration. Vlasyuk and Mitsko (1967) showed in experiments with pea plants that respiration is affected by low fluorine concentrations. A possible explanation for this is that fluorine-induced changes in cell membrane permeability increases access of substrates to sites of oxidase localization. A stimulation of respiration followed by inhibition of photosynthesis has been registered by Belozerova (1964) in leaves of *Bryophyllum* exposed to sodium fluoride. High concentrations of fluorides have been shown to inhibit respiration and low ones to stimulate it in legume seedlings (Applegate et al. 1960). Sublethal fluorine concentrations significantly suppressed the respiration of plum leaves (McNulty and Newman 1956). The same authors (1957) reported an initial increase in respiration intensity in bean and gladiolus exposed to the prolonged effect of fluorine and then a marked decline in respiration after the appearance of heavy chlorosis of leaves. According to Hill et al. (1959) enhanced respiration seems to be more typical of sites adjacent to necrotic mottles on leaves

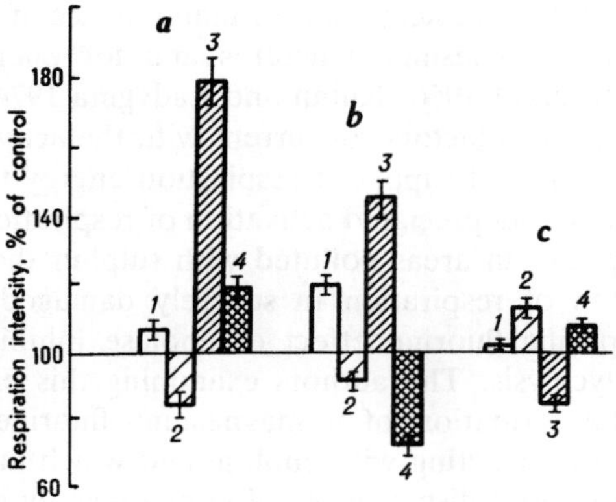

Fig. 21. Changes in respiration intensity in needles of **a** pine **b** larch and **c** spruce following varying degrees of damage by hydrogen fluoride. *1* No visual signs of injury; *2*, *3* and *4* slight, moderate and severe damage respectively

than of any leaf tissue accumulating fluorides. However, this interpretation contradicts the experiments in which a short period or treatment with fluorides was not followed by any visible damage (Applegate et al. 1960).

The effect of fluorine on reduction-oxidation processes in plants depends on such factors as resistance of enzyme systems to fluorine and the presence of compounds diminishing chain oxidation processes. Fluorine, as a source of free radicals, should lead to enhancement of respiration and oxidation reactions. It is also known that fluorides inhibit the glycolytic pathway of respiration while they fail to inhibit enzymes of the pentosephosphate pathway. Such intricate relationships can be responsible for ambiguous results obtained by many workers on the effect of fluorine on respiration.

At the beginning of our experimental fumigation with hydrogen fluoride when there were no signs of visible damage, the intensity of needle respiration increased up to 120% of the control value in larch, 108% in pine but was close to the control value in spruce (Fig. 21). The initial stimulation of respiration in plants affected by hydrogen fluoride is one of the first responses of the organism and is likely to be caused by enhancement of oxidative processes resulting from disturbances in cell membrane permeability and changes in the ion status of the cell (Ilkun 1978; Nikolaevsky 1979). This assumption is consistent with our results which demonstrated a significant activation of peroxidase from the very start of fumigation (see Fig. 20). With the appearance of visible needle damage the intensity of respiration varied between species. It decreased a little in larch and pine following slight damage of needles; this correlates with a drop in peroxidase activity under such conditions. Further damage (40% necrosis of needles) is followed by a rise in respiration level by 45% above the control value in larch, and by 80% in pine; this is associated to some extent with the drastic increase in peroxidase activation.

Increased intensity of respiration in plants exposed to deleterious factors is an adaptive defence response since it enables the plant to increase energy expenditure necessary for maintenance of the integrity and functional activity of protoplasmic structures and for energy supply to synthetic processes (Altergot 1966; Rubin and Ladygina 1974). Moreover, the intense effect of injurious factors concurrent with the activation of respiration aggravates the insufficient supply of respiration energy to other metabolic processes. Such energy-depreciated activation of respiration was observed in some deciduous species in areas polluted with sulphur dioxide (Tarabrin 1974). The inhibition of respiration in severely damaged larch is probably related to the specific fluorine effect of enolase inhibition and consequent cessation of glycolysis. The authors examining this process in cultivated plants suggest the formation of a magnesium–fluorine–phosphate complex by fluoride ions inteacting with enolase and which inactivates enzyme molecules. Also, profound disturbances of enolase-catalysed reactions inhibit respiratory gas exchange (McCune et al. 1964; Veinstein and Soldatenkov 1965; McCune and Weinstein 1971). Respiration in pine is inhibited after a longer period following the appearance of extensive necrosis than in larch. The respiration level of spruce fumigated with hydrogen fluoride exceeds the control value by not more than 10–15% even though the content of fluorine in the needles assayed was found to be 4.3 times greater than in the control. No relation was found between the respiration intensity and peroxidase activity in spruce.

The content of ascorbic acid, one of the strongest antioxidants, can illustrate the reducing properties of fumigated conifers. Plants contain a considerable amount of ascorbic acid, especially in their chloroplasts (Franke and Heber 1964) where it acts as a powerful reducing agent performing particular functions in photosynthesis (Rudolph and Bukatsch 1966). According to some authors ascorbic acid plays a conspicuous role as a detoxifying agent in the plant organism injured by acid gases (Nikolaevsky 1965; Keller and Schwager 1977). The positive effect of ascorbic acid under exposure to acid gases is associated with an inhibition of free radical formation and the depression of oxidative processes in plants in the light.

Ascorbic acid was rapidly oxidized in all three coniferous species from the onset of hydrogen fluoride exposure; its hydrogen is used in reduction processes and therefore the content of the acid drops sharply with the appearance of necrosis (slight damage). In the experiment ascorbic acid content was 51, 55 and 62% in pine, larch and spruce respectively (Fig. 22). The decrease in ascorbic acid concentration with simultaneous increase in peroxidase activity was also observed in spruce needles before the appearance of visible symptoms of damage by sulphur dioxide (Keller 1981). The amount of ascorbic acid tends to increase with increasing damage of needles and accumulation of fluorine. Thus, in larch needles, the level of ascorbic acid rises only slightly and only after severe damage. The trees of this species experience an acceleration of oxidative processes and a deterioration of reducing properties; the level of ascorbic acid remains low compared to the control but stimulation of peroxidase is rather high. Distinct accumula-

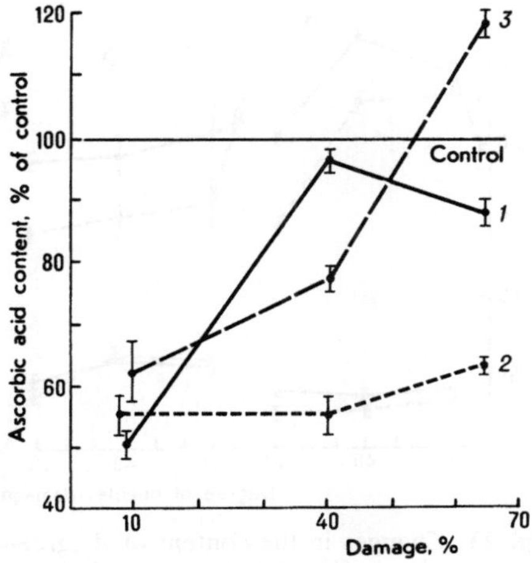

Fig. 22. Changes in the content of ascorbic acid in needles of pine (*1*), larch (*2*) and spruce (*3*) following varying degrees of damage by hydrogen fluoride

tion of ascorbic acid was recorded in moderately damaged pine and spruce with an increase in content by 45 and 15% respectively. With severe damage of the spruce crown the amount of ascorbic acid was 20% above the control value, while in pine it decreased somewhat. Presumably the development of oxidative processes was slightly limited by the enhanced reduction potential resulting from the accumulation of the antioxidant (ascorbic acid). In other words, the plant organism generates defence responses to unfavourable factors, one of which is the accumulation of ascorbic acid.

3.3.2 Dynamics of Photosynthetic Pigments and Reducing Sugars in Needles of Damaged Trees

The harmful effect of fluorides on the photosynthetic apparatus of plants is associated with inhibition of early stages of pigment synthesis, disintegration of chlorophyll and carotenoids and destruction of chloroplasts (Solberg and Adams 1956; McNulty and Newman 1961; Ilkun 1978). The content of pigment is known to be a kind of measure of potential photosynthetic ability of plants; therefore, dynamics of pigments under different degrees of damage may characterize the level of photosynthetic depression. Variations in the content of carotenoids and the sum of chlorophyll a and b were measured in our experiments under conditions of differing extents of damage of the assimilatory apparatus, ranging from insidious to severe damage. We also endeavoured to distinguish the resistance of coniferous species to the destructive effect of hydrogen fluoride on pigments.

The content of chlorophyll in larch was a little above the control value at the onset of fumigation when there were no visible signs of injury (Fig.

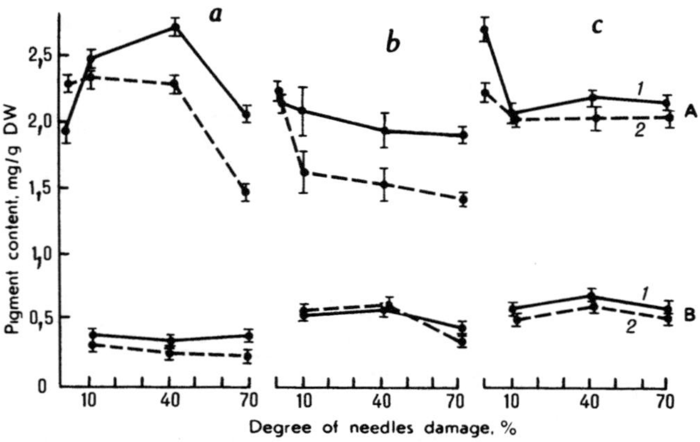

Fig. 23. Changes in the content of *A*, green pigments and *B*, carotenoids, in needles of pine (*a*), larch (*b*) and spruce (*c*) following varying degrees of damage by hydrogen fluoride. *1* Control; *2* experimental value

23). This was apparently the result of defence responses of the tree to the factor causing a decline in photosynthetic ability. An appreciable drop (by 28%) in green pigment concentration occurred at 10% damage of the crown but further increase in needle necrosis resulted in almost no change in pigment content in the rest of the needles whose pigment content was about 75% of the control value. In pine the initial effect of hydrogen fluoride produced a significant increase in the amount of green pigment. Chlorophyll content did not drop significantly with 10% damage of needles and only in severely damaged trees was the content of pigment in the remaining needles 30% less than in the control. The initial response of spruce to hydrogen fluoride involved an insignificant fall in the content of green pigments and with increasing damage of the crown the remaining needles did not differ from the control in their pigment content.

The content of carotenoids in needles injured by hydrogen fluoride showed less change than the chlorophyll concentration (see Fig. 23). Larch exhibited a 16% decline in the content of yellow pigments and only in the case of severe needle damage. Pine and spruce displayed a decreased level of carotenoid content even with slight necrosis of the needles. The maximum difference observed in the amount of these pigments between control and experimental plants with a severe extent of damage was 36% in pine and 15% in spruce. Carotenoids are reactive, unsaturated molecules and can act as antioxidants preventing photodestruction of chlorophyll (Tarchevsky 1977).

The content of chlorophyll in the remaining green needles does not correlate with the degree of crown damage. Dieback of needles proceeds gradually, depending on their resistance to fluorine; the process of pigment destruction and chlorosis occurs as soon as lethal concetrations of fluorine are reached in the photosynthesizing cells of the mesophyll. Under exposure to acid gases the level of photosynthesis may decline not only by destruction of chlorophyll-containing apparatus but also by malfunction of a number of enzymes involved in transformation of sugars and phosphosugars.

A survey of the literature shows that the effect of acid pollutants on carbohydrate metabolism in plants largely depends on the type of pollutant gas and its concentration. Thus, low concentrations of chlorine lead to starch hydrolysis and accumulation of mono- and disugars while high concentrations of chlorine and sulphur dioxide enhance starch content and decrease the amount of reducing sugars (Ilkun 1971a; Spakhova and Ryazantseva 1978). Sitnikova (1964) reported an increase in the content of sugars in polluted woody plants. In needles of common spruce affected by low concentrations of sulphur dioxide the content of starch and sucrose decreases and the content of glucose and fructose increases (Ryšková 1982; Magel and Ziegler 1986). Negrutskaya et al. (1981) examined needles of common pine damaged to differing extents by sulphur dioxide and recorded a rise in starch and drop in soluble sugars with decreasing distance from the source of emission. An increased content of reducing sugars and decline in sucrose level were reported in the leaves of soybean injured by hydrogen fluoride (Yang and Miller 1963); no changes in starch content were found. Other workers (Zwiazek and Shay 1988c) reported a decrease in starch and sucrose levels and an almost two-fold increase in glucose content in pine seedlings treated with sodium fluoride. Ultrastructural studies of leaves of mulberry growing at the aluminium plant showed a very low number of starch grains in the leaves (Torgasheva 1984). The author believes that this result is either from depression of starch synthesis under the influence of fluorine or changes in carbohydrate metabolism which tend to redistribute metabolities and lead to the prevalence of soluble ones. The study of sugar levels in needles of common pine trees that are resistant and sensitive to fluorine and sulphur dioxide demonstrated a higher content of glucose and fructose in sensitive trees but no differences in the content of sucrose (Mejnartowicz and Zukasiak 1985). Jamrich et al. (1989) investigated disruption of metabolism in leaves of willow polluted with fluorine and inferred that the level of damage and destruction of the pigment system in leaves correlated with the extent of disorder in sugar metabolism.

Our determination of the content of reducing sugars in needles of pine, larch and spruce under differing degrees of injury by fluorine show, to a certain extent, disorders in carbohydrate metabolism in these plants. It was found that the content of reducing sugars in all species with any degree of damage were lower than in the control (Fig. 24). However, changes in the level of sugars were shown to vary between species. In larch the sugar concentration dropped by 20% under conditions of slight damage, approached the control value with moderate damage and with severe damage the remaining needles contained 45% less sugar than the control. In pine there was a different correlation between sugar concentration and

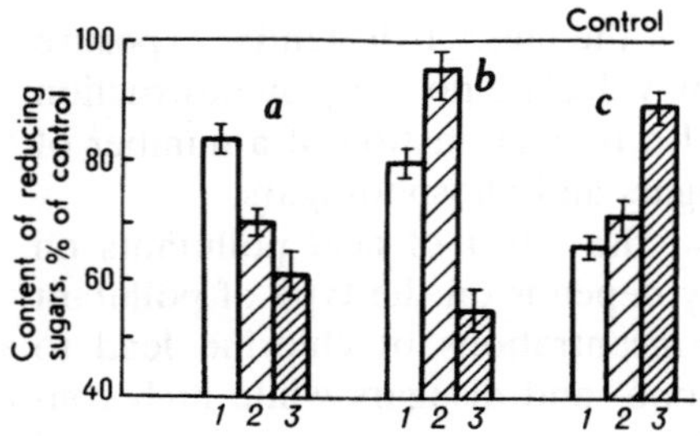

Fig. 24. Changes in the content of reducing sugars in needles of **a** pine, **b** larch and **c** spruce following slight (*1*), moderate (*2*) and severe (*3*) damage by hydrogen fluoride

degree of tree damage: the sugar content decreased with increasing needle necrosis. In spruce slight damage caused a drastic drop in the sugar content (34% of the control) and heavier damage was followed by sugar accumulation, though the content remained lower than in the control. Variations in the content of reducing sugars can be treated on the one hand as a result of photosynthetic and respiratory disorders and on the other hand as specific inhibition of enzyme systems that catalyse transformation of carbohydrates.

The decline in assimilatory activity of the tree leads to a decrease in entry of photo-assimilates into the roots of the damaged plant and to a decrease in root absorptive and biosynthetic activities which, in turn, has a deleterious influence on the rate of xylem flow (Girs 1982). Our evidence suggests that the initial reduction in intensity of upward flow in conifers affected by hydrogen fluoride is not related to the absorption activity of roots, which is rather high even in severely damaged trees.

Observation of the change in water content of hydrogen fluoride fumigated trees was followed by parallel measurements of the content of pigments and sugars. A distinct tendency towards a decrease in moisture before the appearance of visible symptoms of damage has been revealed in all species (Fig. 25). The content of fluorine in the needles was 2.5, 1.8 and 1.5 times above the control value in larch, pine and spruce respectively (see Table 14). Needles lose their water at the same time as damage and fluorine accumulation occur. Thus, the initial decrease in water content is insignificant but it rises dramatically in pine and spruce after as little as 10% damage to the needles (see Fig. 25). In larch the common tendency of conifers to lose water is expressed as a regular decline in water content of the needles according to the extent of their damage.

3.3.3 The Role of Mineral Cations and Organic Acids in Regulation of Ion Balance in Polluted Needles

A mechanism exists for the regulation of cation–anion balance and its maintenance at a certain level in the cell and in the whole plant. This mechanism can be destroyed by alteration to environmental factors and

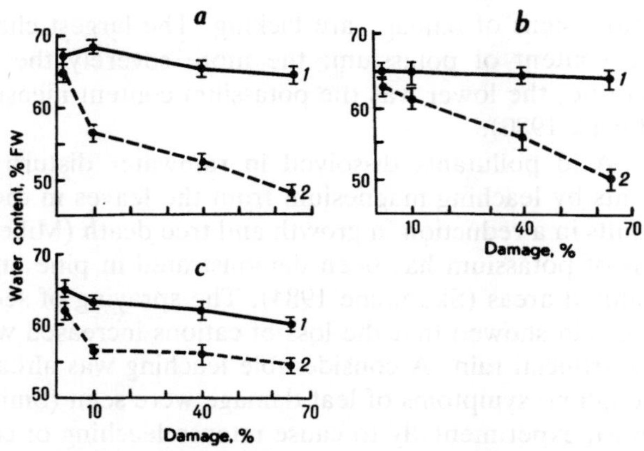

Fig. 25. Changes in the water content in needles of **a** pine, **b** larch and **c** spruce following varying degrees of damage by hydrogen fluoride. *1* Control; *2* experimental value

especially by unbalanced uptake of ions by the plant. With entry of low concentrations of inorganic pollutants the buffer system of the cell sap is "triggered" and hence pH does not change; with moderate and high concentrations the buffer system is unable to completely neutralize the toxic ions and, as a result, there is a pH change (Durmishidze et al. 1981). In leaves, acid gases usually induce a pH shift towards increased acidity whilst ammonia causes a shift towards alkalinity. Changes in pH are registered first in conducting vessels and adjacent cells and then in mesophyll cells (Ilkun et al. 1971).

One of the possible ways to elevate the buffer capacity of the cell sap under conditions of high concentrations of toxic ions is to accelerate the entry to leaves of mineral cations from other organs and to increase cation uptake from the soil. Different plants use different cations to neutralize an excess of anions. Mostly, ions of calcium and potassium in combination with magnesium and sodium are used but rarely sodium with magnesium (Ilkun 1978). The more cells and organs are saturated with free mineral cations, the later irreversible changes occur. The composition and level of cations depend on the individual plant (Ilkun 1971b) and on edaphic conditions of growth (Jurat et al. 1986). Particularly threatening is the effect of acid gases on conifers growing in acid soils. In this case because of an elevated concentration of phytotoxic aluminium in the soil solution the performance of the root system is disturbed and symptoms of so-called aluminium toxicosis can be seen (Weiger 1983); aluminium inflicts lethal damage on thin roots and inhibits absorption of calcium and magnesium ions, which also results in a deficiency of these elements in the needles (Hüttermann and Ulrich 1984; Jurat et al. 1986). Artificial fumigation of maize and pea plants with hydrogen fluoride and sulphur dioxide for 1–3h disrupts the rate of anion and cation uptake from the nutrient solution by roots even when visible signs of damage are lacking. The largest changes were recorded for the content of potassium: the more severely the plants were injured by fluorine, the lower was the potassium content measured in them (Ilkun and Motruk 1980).

Acid pollutants dissolved in rainwater disturb the buffer capacity of plants by leaching magnesium from the leaves in the zone of acid rain; this results in a reduction in growth and tree death (Miller 1984). An appreciable loss of potassium has been demonstrated in pine and spruce in industrially polluted areas (Skuodene 1984). The spraying of seedlings with an artificial acid rain showed that the loss of cations increased with increasing acidity of the artificial rain. A considerable leaching was already observed at pH 4.0, though no symptoms of leaf damage were seen (Smith 1985). Aerosols were shown experimentally to cause intense leaching of cations from needles and leaves (Skiba et al. 1986). A method of "bioindication" of sulphur dioxide and fluoride effects on conifers, the turbidity test, is based on the capacity of pollutants to leach ions from the needles (Hartel and Fuchshofer 1987).

Organic acids play an essential role in the regulation of cation-anion balance, especially when unbalanced ions enter the plant via the root system (Ratner and Akimochkina 1949; Shakhov 1956 etc.). Data on the contribution of organic acids to this balance in the case of ions entering the plant òrganism via the leaves from polluted air are scarce (Yang and Miller 1963; Ilkum et al. 1971; Ilkun 1978). The authors mentioned are inclined to think that organic acids are essential for regulation of the cation-anion balance only until the appearance of visible signs of the toxic effect or during slight visible damage.

In considering the detoxifying capacity of conifers it is necessary to think about specific traits responsible for their elevated sensitivity to acid gases. The majority of conifers are characterized by low buffer capacity of the needles (Pylypes and Redmann 1984) and high acidity of the cell sap owing to a large quantity of organic acids present in the free state. The pH value of the cell sap of needles is 3.6 in common pine, 4.5 in Siberian spruce and 3.6 in Siberian larch; the acidity of young needles early in the growing season is particularly high (Sudachkova 1977). This relates largely to the response of different conifers to acid pollutants.

We have analysed the content of organic acids and mineral cations in the tissues of needles of larch, pine and spruce under varying degrees of damage (Table 15). In spruce and larch an increase in crown damage is followed by a decline in the total amount of cations in the needles and in pine there is a tendency towards a decrease only in monovalent cations with increasing damage. As the concentration of mineral cations in damaged trees fails to increase (except for calcium ions in pine) there is no enhancement of cation influx into the needles. The cations may be lost by being intensively excreted from the root system. As shown by Ivanov (1973) the increase of root exudates in response to deteriorating environmental conditions is not followed by enhanced absorption of cations.

Table 15. The content of free cations in the needles of 10-year-old trees (% DW)

Degree of damage	Na^+	K^+	Ca^{2+}	Mg^{2+}	Sum
Larch					
Control	0.031 ± 0.002	0.47 ± 0.02	0.65 ± 0.03	0.22 ± 0.01	1.43 ± 0.07
Slight	0.025 ± 0.002	0.47 ± 0.02	0.54 ± 0.02	0.15 ± 0.02	1.19 ± 0.03
Moderate	0.013 ± 0.001	0.46 ± 0.03	0.51 ± 0.02	0.16 ± 0.02	1.14 ± 0.02
Severe	0.010 ± 0.002	0.44 ± 0.03	0.44 ± 0.02	0.13 ± 0.03	1.02 ± 0.03
Spruce					
Control	0.015 ± 0.002	0.63 ± 0.03	0.82 ± 0.02	0.13 ± 0.01	1.60 ± 0.06
Slight	0.017 ± 0.001	0.47 ± 0.02	0.67 ± 0.03	0.10 ± 0.01	1.26 ± 0.02
Moderate	0.015 ± 0.002	0.44 ± 0.01	0.65 ± 0.03	0.09 ± 0.01	1.20 ± 0.02
Severe	0.013 ± 0.001	0.45 ± 0.01	0.60 ± 0.01	0.09 ± 0.01	1.15 ± 0.02
Pine					
Control	0.025 ± 0.001	0.56 ± 0.02	0.56 ± 0.01	0.12 ± 0.01	1.27 ± 0.06
Slight	0.020 ± 0.002	0.55 ± 0.02	0.69 ± 0.02	0.12 ± 0.01	1.38 ± 0.08
Moderate	0.022 ± 0.002	0.52 ± 0.02	0.60 ± 0.02	0.12 ± 0.01	1.26 ± 0.05
Severe	0.017 ± 0.001	0.48 ± 0.02	0.63 ± 0.03	0.11 ± 0.01	1.24 ± 0.02

Under experimental conditions fluorine induced a drop in the amount of cations in all three species considered, though to a varying extent. In severely damaged larch needles the sodium content was three times less than in the control, the magnesium content also declined considerably as did the calcium content (more than 30% decrease) but the level of potassium remained practically unchanged. In hydrogen fluoride fumigated spruce trees there were similar changes in the content of potassium, calcium and magnesium but the content of sodium, unlike in larch, failed to vary. In pine the concentrations of magnesium remained at the same level following any extent of tree damage but with increasing necrosis throughout the crown there was a decrease in sodium and potassium accompanied by an elevated content of calcium ions.

In needles slightly damaged by fluorine the content of total organic acids declined, especially in larch (by 37.5%) and pine (by 24.2%) but the decline was insignificant in spruce (by 5.8%). Organic acids accumulated in all species with increasing damage (Fig. 26). There was significant variation between conifers in this process: a direct relationship between needle damage and acid accumulation could be observed in larch (in severely damaged needles organic acids exceeded the control value by 27%); in pine a sharp rise in the level of organic acids followed only after heavy damage to the crown (9% above the control value); and in spruce the accumulation of organic acids reached its maximum in moderately damaged needles (33.6% above the control) but then declined, though it remained above the control value.

In summary, the regulatory action of organic acids in maintenance of the cation-anion balance is displayed when 40% of the crown is damaged in

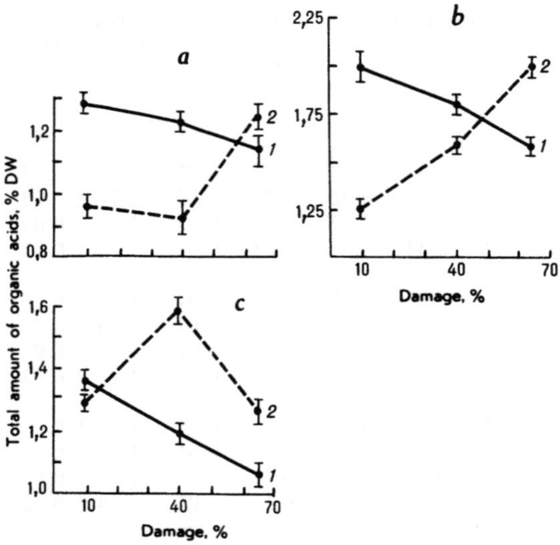

Fig. 26. Changes in the content of organic acids in needles of **a** pine, **b** larch and **c** spruce following varying degrees of damage by hydrogen fluoride. *1* Control; *2* experimental value

pine and 10% in larch. Subsequently the acids enhance an ionic imbalance towards acidification, as also happens when fluorine is accumulated. Our experiments revealed the regulatory action of organic acids in the initial period of spruce needle damage (up to 10%) and in severely damaged needles (up to 70%).

3.3.4 Dynamics of Nitrogenous Substances in Needles of Damaged Trees

Nitrogen metabolism is one of the most important indicators of the physiological state of plants as well as of their adaptive potential to the impact of unfavourable factors and conditions. The importance of certain protein fractions and some amino acids in enhancing plant resistance under extreme conditions has been demonstrated by investigating the metabolism of nitrogen – containing substances (Altergot 1969; Rubin and Artsikhovskaya 1960; Sudachkova 1977; Novitskaya and Chikina, 1980; Genkel 1982). Nitrogen is present in organs and tissues of conifers mainly in the form of protein; at certain periods proteins may constitute 93–99% of the total nitrogen content and nonprotein nitrogen usually does not exceed 15% in the needles of spruce and 20% in pine (Sudachkova 1977; Novitskaya and Chikina 1980). The main part of the nonprotein fraction of nitrogen in conifers is composed of free amino acids and ammonia in the mineral form (Prokushkin 1982).

Our particular aim was to analyse changes in the content of nitrogenous substances during conifer damage before the trees pass to the state of

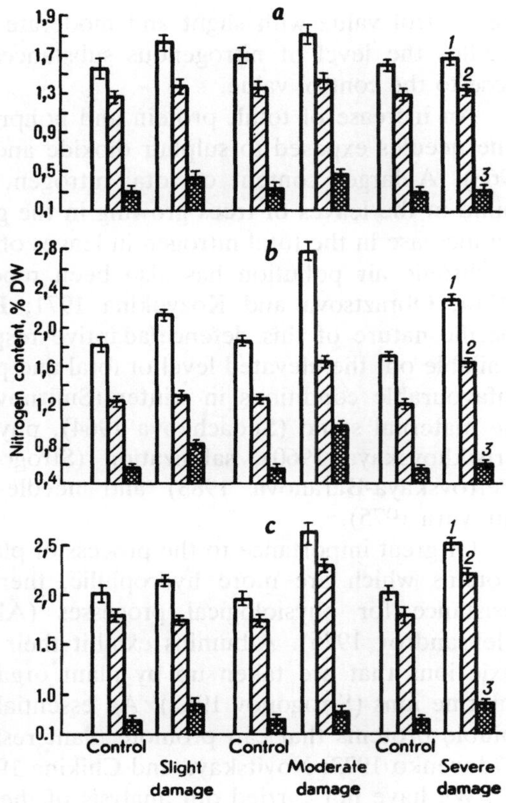

Fig. 27. Changes in needles of **a** pine, **b** larch and **c** spruce in the content of total (*1*), protein (*2*) and non-protein (*3*) nitrogen following varying degrees of damage by hydrogen fluoride

irreversible weakening and decline. The content of total nitrogen was found to increase in the needles of damaged trees, e.g. under slight damage it was 116% in larch, 117% in pine and 107% in spruce as compared to in the control (Fig. 27). With accumulation of fluorine and increase in damage the total nitrogen content increased further, amounting to 155% in larch and 145% in spruce; the content of nitrogen decreased again with severe damage of the needles.

In slightly damaged larch the total nitrogen content increased because of a rise in nonprotein compounds (their amount increased by 45% in the experiment) and in moderately and severely damaged needles it increased because of a rise in both nonprotein and protein compounds. In slightly damaged spruce needles the content of protein nitrogen did not change significantly, while the content of nonprotein nitrogen increased by 68%; in more damaged needles the content of total nitrogen rose because of an increase in both forms of nitrogenous compounds. In pine, unlike in other coniferous species, there was only a little increase in the nitrogen content; this was mainly owing to nonprotein substances which rose by 42–53% of the control value with slight and moderate damage. In severely damaged needles the level of nitrogenous substances dropped somewhat and was close to the control value.

An increase of total, protein and nonprotein nitrogen is also found in pine needles exposed to sulphur dioxide and nitrogen oxides (Negrutskaya 1968). A larger content of total nitrogen, phosphorus and potassium is found in the leaves of trees growing in the grounds of aluminium smelters. An increase in the total nitrogen in leaves of woody plants under conditions of chronic air pollution has also been reported by other authors (Ilkun 1971a; Obraztsova and Kozyukina 1971; Popov et al. 1981). The nonspecific nature of this defence/adaptive response is indicated by the data available on: the elevated level of total and protein nitrogen in leaves under unfavourable conditions in winter (Smirnova 1968), plant suppression by the maternal stand (Sudachkova 1964), phytopathogen action (Rubin and Artsikhovskaya 1960), salinization (Strogonov 1962), low temperatures (Petrovskaya-Baranova 1983) and needle-eating pests (Pleshanov and Pindyura 1975).

Of great importance to the process of plant resistance are water-soluble proteins which are more hydrophilic, thermostable and provide greater resistance for physiological processes (Altergot and Bukholtsev 1970; Aleksandrov 1975). Albumins exhibit their defence properties by binding toxic ions that are taken up by plant organs as shown by, for instance, chlorine ions (Strogonov 1962). An essential role is also assigned to alkali-soluble proteins that can promote plant resistance to unfavourable factors (Udovenko 1973; Novitskaya and Chikina 1980).

We have not carried out analysis of the fractional composition of proteins in stands polluted with hydrogen fluoride. The results obtained on the fractional composition of proteins in pine exposed to sulphur dioxide (Rozhkov and Vasilieva 1981) confirm that the rise in protein nitrogen of needles affected by acid gases can result from water-soluble and alkali-soluble fractions. The data obtained show the increase of these soluble fractions of proteins in the needles of slightly and moderately damaged trees. The level of total and protein nitrogen also increases. In irreversibly weakened trees, total and protein nitrogen decreases and nonprotein nitrogen increases; the fractional composition of proteins changes with an increase in the proportion of salt-soluble and alcohol-soluble fractions and a decrease in water-soluble and alkali-soluble fractions.

The nonprotein fraction of nitrogen increases under the effect of acid gases primarily because of qualitative and quantitative changes in free amino acids (Yang and Miller 1963; Ignatenko and Tarabrin 1975; Kozyukina 1976). The study of amino acid metabolism of woody plants in industrially polluted areas revealed that resistant species accumulated amino acids in leaves by activating the processes of amino acid synthesis, while nonresistant plants enhanced their content of amino acids by inhibiting synthetic processes and accelerating hydrolytic decomposition of proteins (Tarabrin and Ignatenko 1976; Ignatenko 1981). The conclusion drawn was that accumulation of amino acids in resistant species under conditions of acid gas impact were an adaptive response and the sharp rise in the amino acid content in nonresistant species, along with the appearance of visible damage of leaves, indicated that the biosynthetic processes were out of balance.

3.4 Summary

Fluorine can cause acute, drop-like burn, chronic or latent damage to plants. The acute damage results mostly from a single emission of a high fluorine concentration into the atmosphere during aluminium plant failure. It gives rise to rapid (during several hours or days) development of chlorosis with subsequent necrosis of needles or growing shoots. The mechanism of damage basically involves the photodynamic effect of chlorophyll in an acidified cellular medium. Drop-like burns are formed by the effects of dew drops in which the concentration of fluorine rises to a level causing burns by repeated evaporation of droplets at the same site on the leaf (needles). Chronic damage, often embracing vast areas, poses the greatest threat to forests. It is induced by the long-term effect of relatively low concentrations of hydrogen fluoride.

Being highly aggressive fluorine can penetrate directly into all organs of the woody plant through stomata, the cuticle of young shoots, bark of the trunk and branches, epiblema and secondary bark of the roots. When fluorine enters the plant by another method, as in the experiment, its transport into other tissues and organs results in similar final damage of the tree. Under conditions of constant pollution of the environment with fluorine-containing industrial emissions the level of fluorine concentrations accumulated in living tissues is rather steady. In all conifers the largest concentration of fluorine is observed in needles, is somewhat less in the phloem and least in sapwood and root tissues. These tissues can be arranged in the same order according to the intensity of fluorine movement from them. The largest amount of fluorine is translocated via the phloem and only a small amount moves by the path of upward flow; this restricts the movement of fluorine from the roots to the crown.

Mechanisms of tree resistance to the toxic effect of fluorine as well as to other unfavourable factors are already present in the plant. They involve detoxification by chemical transformations, isolation of fluorine in dead and poorly functioning tissues and its removal into the environment. The main pathway of fluorine removal is through the roots but loss of needles and transpiration also remove fluorine. The site of fluorine isolation is the outer bark, which fluorine enters from the phloem, and to a lesser degree isolation occurs in the dead xylem. For the living tissues of phloem and xylem to function properly it is necessary for them to be constantly renewed by radial growth of wood and bark formation.

The alterations in physiological-biochemical features of conifers that lack visible signs of damage or show only initial symptoms (when the fluorine content in needles is 1.5 to 2.5-fold above that of the control) indicate the function of defence mechanisms aimed at fluorine detoxification. During this period, the detoxifying enzyme peroxidase is activated, the intensity of respiration is enhanced, ascorbic acid is oxidized and the hydrogen produced is used in reduction processes, and the amount of organic acids drops (in addition, the content of free cations increases in pine) to prevent a shift in ion balance towards acidity.

The content of green pigments in the needles initially either increases or decreases a little under the influence of fluorine; then, following intense development of necrosis (up to 40% damaged needles), the amount of chlorophyll in the remaining green needles is kept at a rather high level that is 70–75% of the control value or, in spruce, approaches this value. The pigment destruction leading to chlorosis of the needles does not proceed gradually and only occurs after fluorine accumulation reaches a threshold level. This level, as well as resistance in general, varies not only between different-aged needles but also between similar-aged needles and even between apical and basal regions of individual needles.

Increasing damage, where needle necrosis exceeds 40% or more and the fluorine content in remaining green needles is three to four times above that of the control, shows that most physiological-biochemical parameters indicate a decline in defence properties, enhancement of metabolic disorder and, finally, irreversible weakening of the tree. Though larch still retains a high peroxidase activity but with respiration inhibition and a decline in the content of ascorbic acid this has an adverse effect since it increases the oxidative potential of cells. The accumulation of organic acids and reduction in free cation levels causes an irreversible ion imbalance. Profound disturbances occur in carbohydrate and protein metabolism. From this point the number of needles lacking symptoms of necrosis and still containing chlorophyll decreases rapidly in all conifers; their water content declines with upward flow in the trunk falling to a minimum and the tree begins to dry. Siberian spruce is more resistant to a high level of hydrogen fluoride than are pine and larch; this is confirmed by measurements of the intensity of respiration, the content of pigments, sugars, ascorbic and organic acids as well as the pH of needle cell sap and the water content of needles.

The characteristics of the tree passing to the state of irreversible weakening are very important, both for diagnosis of acute damage and for control of stands situated in the zone of chronic toxic effect. A complete failure to produce new auxiblasts and a reduction in the amount of living brachyblasts by 50% or more may serve as indicators of irreversible weakening in larch. Young spruce trees lose the ability to repair damaged needles on most shoots from the current year and from the two previous years.

Chronically weakened trees decline when 3-year-old needles are absent and 2-year-old shoots are partially dead. Irreversibly weakened pine is characterized by the same features as for spruce. The irreversible weakening of trees of all three species follows the accumulation of fluorine in the needles up to 20 mg% dry weight or more.

Chapter 4
The Role of Insects in Fluorine-Damaged Stands

4.1 The Effect of Fluorine on Insects

A great many works have been published in the last 15 years on the role of phytophagous insects in forests polluted with industrial emissions. Nevertheless, our knowledge of the conditions of development, adaptation and performance of phyllophagous and xylophagous insects (referred to as phyllophages and xylophages) in forests polluted with fluorine-containing emissions is limited. The role of insects in decline of such stands is sometimes undoubtedly overestimated, especially for coniferous forests, and should be distinguished from the capacity of needles to accumulate fluorine during the whole growing period and over many years. Trees can decline in vast areas without any contribution from insects.

It is first necessary to note the diverse effect of pollutants, including acidic ones, on plant-feeding insects; for instance, a comparative analysis of dendrophilous entomofauna of forests in the vicinity of aluminium smelters and pulp and paper plants showed that fluorides had a much larger impact on the composition of xylophage populations in coniferous forests than did sulphur compounds. The differences were particularly evident from the presence and number of longhorn beetles of the genus *Monochamus* and weevils of the genera *Hylobius* and *Pissodes* in polluted stands. The toxic effect of fluorides on insects is exerted mainly through insect feeding but sulphur compounds act to a greater degree as fumigants and contact insecticides. Nevertheless, hydrogen fluoride has a stronger repellent action than sulphur dioxide and physiological-biochemical transformations induced in trees are less attractive to dendrobionts resistant to acid gases.

In the initial phase fluorine has a greater effect on the trees than on the xylophages feeding on them and there is an initial rise in population density of these insects, which differ in the aggression of their attack. A short-term increase in the number of xylophages precedes the reduction in species composition, which starts long before the decline of the stand. Sulphur compounds do not make xylophages migrate from the polluted area; a fluctuation in the insect community then basically follows the change of the stand.

It has been shown experimentally (Mikhailova and Anisimova 1978) that concentrations of fluorine lethal to imagos and larvae of longhorn beetles of the genus *Monochamus* (*M. sutor* L., *M. galloprovincialis* Oliv., *M. urussovi* Fisch.) are tenfold higher than concentrations found in heavily polluted zones around aluminium smelters. In chambers, beetles of these species died only after 5 days or more under hydrogen fluoride concentrations exceeding 100- to 180-fold the maximum concentration near aluminium smelters (up to $30\,mg/m^3$) and larvae died when the fluorine content in the phloem was 50 times higher than in trees growing in the zone of heavy pollution (up to 53 mg% dry weight). In the experiments the first larvae died after 3 days and the last larvae after 14 days, having had enough time to make tunnels 3 cm long in the toxic phloem.

Selikhovkin (1985) followed experimentally the effect of fluorine on development, survival and fecundity of lepidopterous insects by monitoring feeding. To the nutrient medium used for raising caterpillars was added fluorine in concentrations similar to ambient ones. As a result practically all stages of the life cycle deteriorated: mortality increased, the number of caterpillars and pupae decreased, the proportion of pupating caterpillars and emerging butterflies declined, fecundity was reduced and the period of development was prolonged. Fluorine was also shown to have a cumulative effect on phyllophages; thus the mortality of caterpillars inevitably increased by the end of their development. The gypsy moth (*Ocneria dispar* L.), pine moth (*Dendrolimus pini* L.), satin moth (*Leucoma salicis* L.) and rusty-greyish tussock moth (*Pygaera anastomosis* L.) were used in the experiments. The latter two species are encountered in fluorine-polluted forests.

Fluorine can repel many insects. At the same time the attraction of burned-out forests can surpass this repellent effect as shown, for instance, by the case of longhorn beetles of the genus *Monochamus*. The beetles are concentrated in newly burned-out spots in the zone of moderate pollution where they are usually absent or are encountered only occasionally.

4.2 The Reduction in Mechanisms of Plant Resistance to Damage by Insects

Damage of conifers by industrial emissions, the main components of which are hydrogen fluoride and sulphur dioxide, is found to be followed by alterations in carbohydrate and nitrogen metabolism as well as by the production of defence substances. Depending on the degree of disturbance to the physiological processes by fluorine, the resulting changes are either of a defensive nature or are indicative of irreversible weakening of the tree. There is an inverse relationship between the fluorine accumulation in tissues of weakened trees and the content of defence substances in them. Thus, maximal saturation of tissues with fluorine corresponds to strong inhibition of synthesis of resinous substances (Rozhkov and Massel 1982). Resinous substances are the principal means by which the living conifer defends itself against phytophagous insects. The protective function of resin is associated with its chemical and mechanical effects on pests. Resins act as repellents for insect imagos and as toxins to larvae and to a few imagos.

The results of the study of the effect of fluorine on the composition and content of resinous substances in coniferous tissues, carried out in the Laboratory of Wood Plant Pathology of the Siberian Institute of Plant Physiology and Biochemistry (Irkutsk, USSR), are reported elsewhere. The data presented below have been described in detail in the book by Rozhkov and Massel (1982).

As has already been discussed, conifers damaged by toxic gases alter their carbohydrate metabolism and reduce the amount of mono- and disaccharides in the needles and phloem. The synthesis of resinous substances is therefore inhibited by a deficiency of low molecular weight precursors. Analyses indicated that fluorine accumulation exceeding 4–4.5 times the conventional level in the needles of larch and spruce led to the reduction of both resins and neutral substances by 50%, and in pine by 30 and 50% respectively. Considerable accumulation of fluorine in the phloem of larch (up to 5 mg% dry weight) and spruce (up to 7 mg% dry weight) by fluorine movement from the needles along with metabolites resulted in the decline of resinous substances in the phloem by 3-fold in larch and by 2.5-fold in spruce (Table 16). The decline in resistance of these two tree species is related to the depletion of neutral substances in the phloem. In the samples this amounts to not more than 0.8–2.3% dry weight compared to 2.8–3.4% dry weight of the control. At the same time the level of neutral substances in the sapwood of weakened larch trees may exceed the control level by 1.5-fold.

Alteration in the defence chemical system of conifers produces a most distinct effect on the content of the main components of the monoterpene fraction. The transformation of monoterpenes in the needles of larch and spruce causes a decrease in the content of Δ3-carene to 40.3% and 27.7% of the monoterpene fraction respectively compared to a content of 56.4 and 37.1% of the respective controls; there is also an increase in α-pinene by 1.2-fold, β-phellandrene by 2-fold and limonene in spruce by 6-fold. The phloem of these trees differs from the control by a drastic decline in the content of α-pinene and an increase in β-pinene, limonene and β-phellandrene. The amount of Δ3-carene fails to change significantly (Table 17).

Pine trees exposed to the toxic effect of fluorine accumulate it at a slower rate than do larch and spruce trees. The low rate of fluorine accumulation in pine trunk tissues is in line with the changes in resin content. The level of resinous substances in the phloem of damaged pine trees declines little compared to the control. Moreover, this decline is compensated by the increase in neutral substances by up to 4.4%. A typical

Table 16. The content of resinous substances (% DW) in needles, phloem and sapwood of healthy trees and trees damaged by fluorine-containing emissions from aluminium smelters

Substances	Healthy			Weakened[a]					
	Needles	Phloem	Sapwood	First group			Second group		
				Needles	Phloem	Sapwood	Needles	Phloem	Sapwood
Pine									
Resins	11.7	6.2	4.4	8.9	5.1	5.3	5.3	4.5	2.0
Neutral	5.7	3.6	1.5	2.9	4.4	2.2	2.0	1.7	0.8
Acidic	6.0	2.6	2.9	6.0	0.7	3.1	3.3	2.8	1.2
Spruce									
Resins	9.6	7.7	3.6	4.4	3.1	1.2	3.6	2.4	1.0
Neutral	5.1	3.4	1.6	1.7	0.8	0.3	1.5	1.0	0.5
Acidic	4.5	4.3	2.0	2.7	2.3	0.9	2.1	1.4	0.5
Larch									
Resins	4.8	6.9	1.6	2.3	2.1	1.3	1.9	1.6	0.9
Neutral	2.6	2.8	0.7	1.2	0.8	1.0	0.8	0.7	0.3
Acidic	2.2	4.1	0.9	1.1	1.3	0.3	1.1	0.9	0.6

[a] Weakened trees of the first group show resistance to xylophagous colonization and trees of the second group are colonized by xylophages; see also Table 17.

Table 17. The content of monoterpene hydrocarbons (% of fraction) in needles and phloem of healthy trees and trees damaged by fluorine-containing emissions from aluminium smelters

Conditions of trees	Object of analysis	α-Pinene	Camphene	β-Pinene	Δ3-Carene	Myrcene	Lymonene	β-Phellandrene	γ-Terpinene	Terpinolene
Pine										
Healthy	Needles	68.7	3.9	4.1	17.2	3.5	0.8	1.5	Trace	Trace
	Phloem	40.3	0.7	3.1	45.7	1.4	1.5	4.3	Trace	2.7
Weakened										
First group	Needles	34.6	1.2	1.5	53.2	0.8	1.9	1.9	0.5	2.4
	Phloem	38.0	0.5	5.2	50.0	0.5	1.0	1.0	–	3.8
Second group	Needles	54.6	4.8	10.0	10.0	7.7	0.6	7.0	3.5	0.5
	Phloem	70.5	5.3	3.7	Trace	6.8	4.4	6.9	2.0	Trace
Spruce										
Healthy	Needles	27.3	11.6	6.6	37.1	2.8	1.6	6.4	0.9	5.1
	Phloem	74.0	Trace	10.6	2.0	1.0	4.4	6.1	–	Trace
Weakened										
First group	Needles	32.7	6.4	8.9	27.7	3.1	13.2	11.4	Trace	Trace
	Phloem	58.2	Trace	5.3	3.5	2.6	11.6	14.5	–	Trace
Second group	Needles	40.6	1.9	12.3	5.6	4.1	10.6	22.0	0.5	Trace
	Phloem	83.0	1.8	4.9	Trace	1.3	2.9	4.3	1.9	Trace
Larch										
Healthy	Needles	22.7	1.1	5.4	56.4	Trace	2.3	3.6	Trace	2.2
	Phloem	24.5	0.4	13.9	46.0	–	2.7	6.5	1.0	2.5
Weakened										
First group	Needles	24.9	–	17.9	40.3	3.1	2.2	7.7	0.5	2.8
	Phloem	16.3	Trace	20.1	44.2	Trace	8.1	10.3	Trace	1.0
Second group	Needles	43.9	Trace	7.1	36.2	2.9	1.3	4.5	1.7	3.7
	Phloem	58.7	–	5.3	19.8	3.0	2.6	4.2	7.0	7.0

response of pine to fluorine damage is the activation in sapwood of resins by 1.2-fold and of neutral substances by 1.4-fold above the control. The composition of monoterpene hydrocarbons is actively transformed in pine needles, in contrast to larch and spruce: Δ3-carene is accumulated by up to 53.2% against 17.2% in the control and α-pinene shows a 2-fold decrease.

The reduction in resistance of larch, pine and spruce to xylophage colonization is followed by diminished synthesis of resinous substances. Compared to healthy trees the content of resin drops 4.3-, 3- and 1.3-fold in the phloem of fluorine-damaged trees of larch, spruce and pine respectively. Similarly, the loss of resistance is distinctly shown by a decrease in the amount of neutral substances in the phloem by 75% in larch, 70.6% in spruce and 52.8% in pine (see Table 16). Similar changes are also recorded in the sapwood and needles of trees being colonized by xylophages. The content of resins declines by 2.5- and 2.7-fold in the needles of larch and spruce and the content of neutral substances, by 3.2-fold and 3.4-fold. In pine sapwood resins and neutral substances decrease by 50%.

The attack on trees by trunk pests influences the synthesis of monoterpene hydrocarbons. In the phloem of larch trees being colonized by xylophages the content of α-pinene increases by 58.7% and Δ3-carene decreases by 57%, thus impoverishing the composition of essential oils, and this is followed by a threefold increase in the content of myrcene and terpinolene and by a reduction of β-pinene and β-phellandrene (see Table 17). In the phloem of irreversibly weakened pine and spruce trees essential oils are synthesized to form primary α-pinene, yielding 70.5% in pine and 83% in spruce. The Δ3-carene content declines to leave only trace amounts. Such a decrease in Δ3-carene and maximal concentration of α-pinene in the phloem suggests the complete loss of tree resistance to xylophage colonization. In addition, an increased content has been found in camphene (7.5-fold), myrcene (4.8-fold) and limonene (2.9-fold) in pine samples; the amount of β-pinene and β-phellandrene is also somewhat elevated.

The trees pass from the category of weakened though retaining a relative resistance to colonization by xylophages to the category of being available for colonization by decreasing the total content of resins in the phloem by up to 5% or less in pine, by less than 3% in spruce and by less than 2% in larch. In addition, the neutral fraction of resins in the phloem is reduced and the concentration of Δ3-carene in essential oils is decreased by up to 20% in larch and to trace amounts in pine and spruce.

4.3 Succession of Entomocenosis in Polluted Stands, Resistant Groups of Insects and Adaptation

In spite of favourable conditions, in terms of the physiological state of trees for development of dendrophylous insects, especially xylophages, most species become rare in hydrogen fluoride polluted forests and many of them leave altogether. Impoverishment of the fauna proceeds gradually with the accumulation of fluorine in trees, bushes, grasses, forest litter and soil. A too-detailed subdivision of changes in the composition and number of dendrophylous insects during the process of stand decline is not believed to be necessary. We have distinguished three stages of this process: (1) local damage and decline of stands in the vicinity of the source of emission with relatively slight damage of stands in the adjacent area, allowing retention of the population composition of needle-eating and trunk pests close to the original; (2) mass decline of trees with accumulation of fluorine in their tissues and the environment to concentrations that are repellent and toxic to free-living phyllophages with gnawing mouthparts and to most xylophages, thus resulting in migration of insects from the polluted region and a reduction in the numbers of remaining species; and (3) substitution of declined conifer stands for deciduous species and formation of new insect populations with few dendrophilous species.

The first stage lasts from 2–3 to 8–10 years at different sites in the polluted area (within the zone of lethal accumulation of fluorine by plants) depending on the amount of fluorine emitted by the aluminium smelter, the composition and age of stands and topography of the land. The higher the fluorine concentration in emissions, the more rapid is the development of the next stage of the process. The second stage may continue for two decades with continuous impoverishment of the population of dendrobionts. The species which are resistant to fluorine exposure at the beginning of this stage cannot tolerate a further increase in concentration in their food. In the third stage dendrophages of conifers are represented by a few species (basically those with sucking mouthparts) that survive on saplings of pine, larch and spruce growing under the deciduous forest canopy. The successive decrease in populations of dendrophilous insects in coniferous forests polluted with fluorine can also be shown for trunk pests. Horntails are the first to disappear in the zone of high fluorine concentration, the next to disappear are metallic woodborers and longhorn beetles and the last to go are bark beetles. Among bark beetles there are species that are relatively resistant to fluorine at ambient concentrations. They are found in polluted stands until the end of the second stage of the decline when colonies are dense.

The results of investigations carried out over many years by our laboratory on xylophages in healthy, weakened (by divergent factors) and declining larch forests in the Baikal area (Rozhkov and Byalaya 1966; Anisimova and Sokov 1975; Rozhkov et al. 1975; Berezhnykh 1978, etc.) can be used to compare species diversity of these insects in different habitats with those in larch forests exposed to fluorine-containing emissions (Rozhkov 1981). With respect to the species composition of xylophages, only larch forests that are declining after temporary flooding and unbarked timber in timber yards are inferior to larch stands polluted with fluorine,

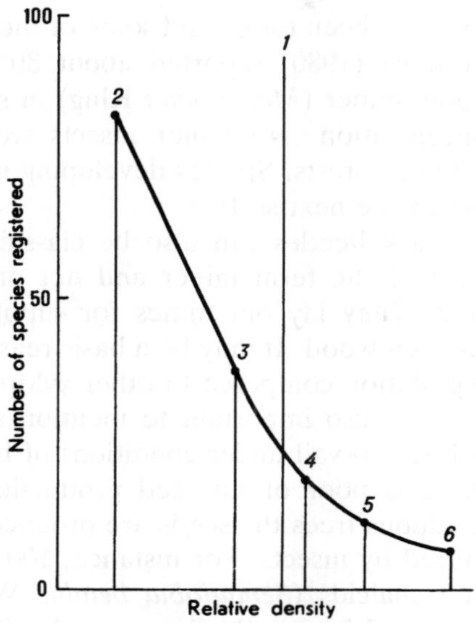

Fig. 28. Species diversity and relative density of populations of abundant xylophagous species developing on larch trees around the Baikal area. *1* Different habitats in general; *2* naturally drying old trees; *3* burned-out forests; *4* weakened and drying stands exposed to fluorine emissions from aluminium smelters; *5* stands weakened and drying due to temporary flooding and silting of the soil; *6* non-peeled timber in timber yards. The number of species is given in absolute values and the density of abundant species population in relative terms.

even at the beginning of the second stage of weakening and decline. In burned-out spots of larch forests the composition of xylophages is twice as rich and in naturally drying old larch forests it is four times richer (Fig. 28).

Among fluorine-resistant phytophages the first widespread group of insects consists of pests with sucking mouthparts: aphids of the subfamily Aphidoidea, tortoise scales of the family Coccidae and plant breeding bugs. Some gallic mites of the family Eriophyidae should also be classified with the resistant sucking phytophages. The mechanism of resistance of the insects mentioned is unknown. It may be relevant that on branches they feed on the sap of young phloem and sapwood cells adjacent to cambium and on leaves they either use the contents of palisade parenchyma cells, avoiding epidermal cells, or specifically suck out the developing cells of the gall parenchyma. The second group of commonly occurring and relatively abundant insects (on deciduous trees) in fluorine-polluted forests is composed of phyllophagous miners with small lepidopterous insects, especially smoky moths (family Gracillariidae), prevailing among them. Caterpillars of many of these moths mine leaves during the whole period of their development and may even pupate in mines. Leaf miners of the subfamily Blannocampinae (family Tenthredinidae) that develop on birch and aspen have also been found and some of them cause mass infestation. Volkova and Vasilieva (1980) reported about 80% leaf damage to trees by the birch pygmy miner (*Messa nana* Klug) in some sites of the zone of high fluorine concentration. No miner insects were found in the needles of trees in polluted forests. Species developing in conifer buds and shoots will be dealt with in the next section.

Bark beetles can also be classified with the miner insects in a broad sense of the term miner and not only larvae but also adult bark beetles mine. They lay out mines for supplementary feeding in the phloem and outer sapwood. It may be a basic reason for higher resistance of bark beetles to pollution compared to other xylophagous insects.

It is also important to mention a group of chalcids in which dipterans (gallfly) prevail under conditions of heavy pollution with hydrogen fluoride. There is poor or no seed production in conifers affected by fluorine; in deciduous trees the seeds are produced more or less regularly and are often injured by insects. For instance, 100% damage of birch seeds by the gallfly seed chalcids (*Semudobia betulae* Winn.) was recorded in large areas of weakened forests (Volkova and Vasilieva 1980).

A particular group of cryptozoic or semi-cryptozoic phyllophages of deciduous species consists of leaf-rolling moths, casebearers, bagworms (lepidoptera) and leaf rollers (coleoptera). The numbers of some of these is high even under conditions of heavy fluorine pollution.

No actual outbreaks of free-living leaf- and needle-eating insects were observed in the zones of severe and moderate fluorine concentrations. Moreover, the number of lepidopterous species of the families Tortricidae, Geometridae and Orgyidae and sawflies of the family Tenthredinidae were found to exist on deciduous trees and they may be abundant in these zones. This can be explained by a combination of many factors, the main ones being: an adequate resistance of these insects to acid gases, a short-term (1 year) cycle of insect development, physiologo-biochemical transformations in the leaves at the stage of weakening of the food plant, which is beneficial for insects, and decline in the activity of parasitic insects and diseases. The physiologo-biochemical transformations that may be beneficial involve, on the one hand, an improvement in the quality of leaf (needle) tissues as food and, on the other hand, a decline in the content of plant defence substances: toxins and repellents. With regard to the decrease in activity of natural mortality factors, regulation was shown to be by the decline in a number of parasitic hymenopterans and by an antiseptic effect of fluorine on viruses and bacteria that could cause infections in hydrogen fluoride polluted areas (Selikhovkin 1982, 1985, 1987).

Analysing the dynamics of density of phyllophage populations on birch and aspen in stands polluted with fluorine and sulphur compounds, Selikhovkin (1985, 1987) inferred that specific insect communities were formed in such stands and that the species composing them (e.g. geometers of the genera *Biston* and *Abraxas* and a weevil of the genus *Apion*) either are preadapted or they develop genotypic adaptations to the pollutant effect. Preadaptation of these insects seems to be of primary importance in this case; such characteristics are acquired during the process of formation of the race, species, genus or ecological group affected by the ambient conditions. Insects with the developed preadaptation adjust to new changes in the environment over several generations.

The normally elevated resistance of some phytophages to natural pollutants contained in food may serve as a physiological preadaptation to life in forests polluted by industrial emissions. Under natural conditions the insects feed on the surface of plants; for example, they skeletonize leaves by gnawing at the epidermis and underlying palisade parenchyma or they feed on lichen and some bagworms (Psychidae) even feed on flakes of bark. Such food is always toxic to some extent as it absorbs natural pollutants from the atmosphere. These pollutants can arise from materials formed by weathering of the earth's crust, forest fires, volcanic eruptions, gases evolved by plants, etc. Insignificant concentrations of carbon monoxide, sulphur dioxide, nitrogen oxides, chlorides, fluorides and mercaptans can be synthesized in the atmosphere. Adaptations of phytophages to them can be considered as a simultaneous preadaptation to higher concentrations of fluorine, sulphur and other toxic compounds from industrial emissions.

In one of the young pine forests within the zone of low concentration of fluorine we found four species of bagworms of the genera *Acanthopsyche*, *Dyplodoma* and *Bacotia*. They were numerous, especially if one bears in mind that there is no predominant species among forest members of this family in Siberia. Presumably, along with a decline in activity of parasites, predators and pathogenic microorganisms, it is resistance to the effect of fluorine that is the basic condition for successful breeding of the bagworms found.

When the atmosphere and food sources are polluted with fluorine, the ecology of insects can undergo essential changes, which also form the basis of preadaptation. It can involve changes in habitat coordination, phenology, character of migration and ethology. Conifer scales of the genus *Physokermes* may serve as an example of a species changing its habitat. They are known to inhabit spruce trees at forest edges, along roadsides, in sites of felled young forests and in other open places. Under high fluorine concentrations conifer scales move into the depth of the forest. This can be accounted for both by thinning of maximally weakened, declining stands and by defence functions of tree crowns. Some other phytophages also move under the forest canopy.

Anisimova and Sokov (1975) followed variations in bark beetles for preference of different food tree species and microhabitats. Thus, the large larch bark beetle (*Ips subelongatus* Motsch.) prefers larch to other tree species under normal conditions but in fluorine-polluted forests it colonizes pine more willingly. Stumps in these forests are more densely inhabited by the large pine beetles (*Blastophagus piniperda* L.) than are drying trees

Table 18. Colonization of microhabitats by the large pine bark beetle in the zone of high fluorine concentration (average number per dm^2)

	Declining trees	Stumps	Windfall trees and logs
Nests	1.3	2.8	2.1
Eggs	57.4	116.0	26.3
Young beetles	15.0	15.0	4.2

(Table 18), though beetle survival (percentage of young beetles to the number of eggs laid) on drying trees is twice as high as on stumps. Windfall trees and stored timber are intermediate in this respect.

No phenological variations have been revealed in coniferous dendrophages. According to Selikhovkin (1985), birch chalcids (*Apion simile* Krby) are characterized by their earlier onset and cessation of supplementary feeding in fluorine-polluted forests; however, the total duration of beetle flights decreased with increasing pollution of the stands. Similar behaviour was observed in the black birch leaf roller (*Deporaus betulae* L.).

The ability of pollution to affect insect migration can be illustrated by the case of the large pine weevil. This pest, which is hazardous for pine sapling growth, avoids breeding in hydrogen fluoride polluted forests even in sites of freshly felled trees. Nevertheless, during supplementary feeding the weevil is concentrated in young pine forests in the zone of high pollution and causes damage to buds and green shoots in their early and middle phases of growth. Then the beetles migrate into non-polluted or only slightly polluted stands to oviposit.

4.4 Survey of Dendrophages of Conifers

Among plant-feeding mites of Siberian coniferous forests polluted by fluorine-containing emissions is the pine coleophyllous mite (*Eriophyes* sp.), a member of the family Eriophyidae, and which is found in high numbers in these forests. This quadruped mite seems to be a newly discovered species. It breeds on severely damaged needles with necrotic apices and the eggs are laid on the inner flat part of a needle at the base. Sometimes adult individuals hibernate along with the eggs. The larvae emerge in May and sucking of the contents of needle cells by both larvae and adult mites accelerates needle drying. The pine coleophyllous mite was not found in pine forests polluted with sulphur dioxide. The larch bud mite (*Eriophyes pini laricis* Nal.), widely encountered in Eastern Siberia, increases its numbers in fluorine-polluted stands but not to the same extent as the pine coleophyllous mite or the gall-forming species of this and the *Phyllocoptes* genera which develop on deciduous trees.

Eleven species of dendrophilous aphids of the two subfamilies Adelgoidea and Aphidoidea, namely *Adelges laricis* Vall, *A. tardus* Dreyf, *A. tardoides* Chol., *Sacchiphantes abietis* L., *Cholodkovskya viridana* Chol., *Schizolachnus pineti*. F., *Cinara kochiana* C.B., *C. pini* L., *C. laricis* Walk., *C. boerneri* H.R.L. and *C. pinea* Mordv., have been recorded in Eastern Siberian coniferous forests polluted with fluorides. Only the last four species can be found in abundant quantities, forming large colonies on shoots and branches in the zone of both moderate and high fluorine concentrations. Chermes (the first five species listed above) are not very numerous and some species are rare; they are more often encountered during the first years after the start of smelter operation.

Spruce bud scales (*Physokermes* spp.) should be ranked with the most abundant pests of spruce in forests polluted by emissions from aluminium smelters. The two to three species have not yet been identified but they occur in large numbers on young trees in the zone of moderate fluorine concentration. In some sites, especially along roadsides, there is total colonization of spruce saplings. They have also been found in the zone of high fluorine concentration and in some of these areas they are quite common. One species has been found on the tops of 40- to 60-year-old damaged trees at a height of 8–10 m, as well as on saplings and the lower branches of adult trees.

The composition and number of coleopterous insects has been analysed over many years in coniferous stands during the different stages of pollution with fluoride in studies by Anisimova and Sokov (1975) and Anisimova (1980, 1981, 1985), research workers in our laboratory. Of particular interest are data on the period of intense reduction in the fauna composition and the density of populations of species susceptible to the impact of fluorine. In the period of 6–8 years after the start of plant operation 42 species were recorded to occur in the zone of high fluorine concentration (the list includes four species of click beetles that damage roots of young trees and breed in the rotting wood of conifers). The species recorded were:

Powder-post beetles (Bostrychidae)
Grooved powder-post beetle (*Stephanopachys linearis* Kug.)

Click beetles (Elateridae)
Coniferous click beetle (*Adelocera fasciata* L.)
Mottles click beetle (*A. conspersa* Gyll.)
Dark-black click beetle (*Elater pallipes* Kr.)
Bright click beetle (*Selatosomus aeneus* L.)

Metallic woodborers (Buprestidae)
Siberian conifer borer (*Ancylocheira sibirica* Fleisch.)
Taiga conifer borer (*A. strigosa* Gebl.)

Four spotted antaxia (*Antaxia quadrimaculata* L.)
Dark blue pine borer (*Phaenops cianea* F.)
Ridged bronze borer (*Chrysobothris chrysostigma* L.)

Longhorn beetles (Cerambicidae)
Ridged rhagium (*Rhagium inquisitor* L.)
Bright prothorax spruce sawfly (*Tetropium castaneum* L.)
Greyish-brown pine longhorn beetle (*Criocephalus rusticus* L.)
Bronze flat sawfly (*Callidium coriaceum* Payk.)
Violet flat sawfly (*C. violaceum* L.)
Black ridged sawfly (*Asemum striatum* L.)
Larch longhorn beetle (*Xylotrechus altaicus* Gebl.)
Grey longhorn beetle (*Acanthocinus aedilis* L.)
Siberian grey longhorn beetle (*A. carinulatus* Gebl.)
Black fir sawyer beetle (*Monochamus urussovi* Fisch.)
Black spruce longhorn beetle (*M. sutor* L.)
Black pine longhorn beetle (*M. galloprovincialis* Ol.)

Weevils (Curculionidae)
Large pine weevil (*Hylobius abietis* L.)
White-spotted weevil (*H. albosparsus* Boh.)
Pine weevil (*Pissodes pini* L.)

Bark beetles (Ipidae)
Large pine beetle (*Blastophagus piniperda* L.)
Small pine beetle (*B. minor* Hart.)
Black-brown beetle (*Hylurgops glabratus* Zett.)
Fluffy polygraph (*Polygraphus polugraphus* L.)
Teploukhov beetle (*Carphoborus teplouchovi* Spess.)
Spruce tiny beetle (*Crypturgus pusillus* Gyll.)
Black pine beetle (*Hylastes ater* Payk.)
Opaque pine beetle (*H. opacus* Er.)
Conifer wood beetle (*Trypodendron lineatum* Ol.)
Six-teeth bark beetle (*Ips sexdentatus* Boern.)
Typograph bark beetle (*I. typographus* L.)
Large larch bark beetle (*I. subelongatus* Motsch.)
Duplicate bark beetle (*I. duplicatus* R. Sahlb.)
Bark beetle of fires (*Orthotomicus suturalis* Gyll.)
Small larch bark beetle (*O. laricis* F.)
Windfall bark beetle (*O. proximus* Eichh.)
Common engraver beetle (*Pityogenes chalcographus* L.)

The composition of dendrophilous coleoptera on conifers in the zone of moderate fluorine concentration includes a few more species but otherwise it remains very similar to that in the zone of slight pollution.

Of dendrophilous dipterans we have found only gallflies (Cecidomyidae) in coniferous forests polluted with hydrogen fluoride. The larch bud gallfly

(*Dasyneura laricis* Lw.) inflicts severe damage on weakened trees in some places and is more frequently encountered in the zone of moderate fluorine concentration. The red pine gallfly (*Thecodiplosis brachyntera* Schwaeg) is found on young pine trees. Its larvae develop between young needles, which become shortened and thickened at the base. However, this gallfly is rarely encountered.

Plant-feeding hymenopterans are very poorly represented in fluorine-polluted coniferous stands. Sawflies (Pamphiliidae, Diprionidae, Tenthredinidae) can be found there in the first years of pollution; then they disappear or became extremely rare. Horntails (Siricidae) remain there longer. According to Anisimova and Sokov (1975), after 8 years of operation of the aluminium smelter the trees in the zone of high fluorine concentration were still inhabited by the dark-blue pine horntail (*Sirex juvencus* L.), the violet conifer horntail (*S. noctilio* F.) and the black-blue conifer horntail (*S. ermak* Sem.) but all of these were rarely encountered. There is evidence of a decline in the density of the population of ants of the genus *Formica* under the effect of fluorine pollutants (Selikhovkin 1985). The decline in number of parasitic hymenopterans was stated above.

In the middle taiga of Eastern Siberia commonly found leaf roller moths (of the family Tortricidae) which mine buds, young shoots and branches of conifers in the zones of high and moderate fluorine concentration include the median bud needle miner (*Blastesthia turionella* L.), the reddish pine shoot moth (*Rhyacionia buoliana* Den et Schiff.), the larch gall moth (*Laspeyresia millenniana* Adamczewski) and the larch wood moth (*L. zebeana* Saxesen). The median bud needle miner is more common on pine regrowth in the zone of moderate fluorine concentration; the gall moth occurs in weakened young larch trees more often than the wood moth.

The pine weevil (*Petrova resinella* L.) is widely found at the initial period of fluorine pollution of pine forests but remains for longer only in the sites of low accumulation of fluorine in the phloem. In addition to the pests of buds, shoots and branches already mentioned we have found the larch shoot moth (*Blastotera laevigatella* H.-S.) of the family Argyresthiidae and the spruce cone pyralid moth (*Dioryctria abietella* Schiff.) of the family Pyralidae in fluorine-polluted stands; neither species is abundant.

The family of Coleophoridae is represented by the Dahurian casebearer (*Coleophora dahurica* Flkv.) in polluted forests. In Eastern Siberia this phyllophage can occur in large numbers during mass breeding in dusty and polluted urban plantings of larch trees along highways and around industrial areas where acid pollutants are emitted into the atmosphere. In 1987 we even encountered the casebearer in the zone of high fluorine concentration and it was present in high numbers in the zone of moderate fluorine concentration.

No fluorine-resistant free-living lepidopterous species have been found on conifers. They are absent from forests exposed to fluorine-containing emissions for 5–8 years or are only rarely met. However, these insects are found on deciduous trees and are represented by a few species of budmoths, geometers and tussock moths.

It is interesting to mention the appearance of many Siberian acridians (*Gomphocerus sibiricus* L.) during 2 consecutive years in the zone of high fluorine concentration. The Siberian acridian was found in the grounds of the aluminium plant, aggregating in lawns between industrial buildings of the plant. Fluorine is likely to stimulate the propagation of these insects, either by enhancing their fecundity or by reducing the activity of mortality factors.

4.5 Numbers and Outbreaks of Xylophages

The specific composition and numbers of xylophages in coniferous forests polluted with fluorine are determined by the characteristics of tree decline. The process of tree decline proceeds rather rapidly at windward sites in the zone of high fluorine concentration and, accordingly, the role of xylophages in accelerating this process then becomes more important. However, even when present in large numbers the insects cannot determine the fate of the trees; this is predetermined by pollutant toxicosis. The decline of conifers in separate sites of the zone of high fluorine concentration continues for many years. Even trees which retain only some branches with needles and lack resistance to pests can remain viable here for a long time. It is fluorine accumulation in the tissues of the trunk, branches and needles that fulfils the defence function against a majority of xylophages and phyllophages. Field studies over many years on xylophages in coniferous forests that are declining because of emissions from one of the largest aluminium smelters in Siberia provide the dynamics of the composition and number of these insects over a 20-year period. The work has been performed by our laboratory and by the Leningrad Forestry Engineering Academy.

Three stages of transformation of xylophage entomocenosis have been distinguished in fluorine-polluted areas prior to substitution of declined coniferous stands by deciduous species: (1) initial differentiation of the population density of insect species that vary in aggressiveness; (2) successive reduction in the population density and the disappearance of species failing to withstand the effect of fluorine accumulated in the food source; and (3) formation and retention of a relatively steady composition of the most fluorine-resistant xylophages until complete decline of the irreversibly weakened stands occurs.

In the most polluted stands and, in particular, on neighbouring windward slopes, some species undoubtedly vanished even within the first year of smelter operation. The depletion of fauna progressed in successive years and after 5–6 years the composition of xylophages was reduced by more than twofold within the zone of high fluorine concentration. The list of longhorn beetles, bark beetles, borers, weevils and horntails given earlier corresponds to this period. As already mentioned, the horntails are the first to vanish in the zone of high fluorine concentration, borers and longhorn beetles are the next and bark beetles the last to disappear. However, the number of some xylophages first increases and they then disappear later.

The initial promotion of the development of xylophages under conditions of air pollution and the ultimate unacceptibility of pollution by insects can be illustrated by the rise in numbers of the larch longhorn beetles (*X. altaicus* Gebl.) during the first years of pollution and their subsequent disappearance following fluorine accumulation in the tree bark to a maximal level. The adult larch longhorn beetle does not feed; it develops on weakened but viable larch trees (it does not inhabit drying trees) and therefore, weakening of trees is initially favourable for its breeding. However, a gradual increase in the content of fluorine in the inner part of the outer bark, adjacent to the phloem, where the first-stage larvae develop during this most vulnerable period, induces a reduction in the numbers of the beetle and its migration from the zone of high fluorine concentration. The longhorn beetle occurred sporadically in the stands prior to the start of smelter operation. It inhabits old trees. The number of beetles under conditions of initial pollution increased rather rapidly, allowing for its 2-year cycle of development. It was classified with the many species present after 7–8 years by Anisimova and Sokov (1975). According to Kondratov, a forest pathologist (pers. comm. 1987), the longhorn beetle had, by that time, reached its highest numbers in the old larch forest occupying a large area several kilometres eastwards of the plant. This larch forest was cut in the 9th year, following the onset of impact of fluorine emission. No larch longhorn beetles were encountered within the zone of high fluorine concentration in subsequent years.

The top bark beetle (*I. acuminatus* Gyll.), the common engraver (*P. chalcographus* L.) and other species, including weevils of the genera *Pissodes* and *Magdalis*, were the first to decrease in numbers and then to disappear from the zone of high fluorine concentration. The similarity between this phenomenon and the migration of xylophages that develop on coniferous regrowth where large numbers of saplings and polewood are available can be accounted for only by analogous biochemical transformations and the repellent effect of fluorine that is quickly accumulated in the thin bark. Along with the migration of many xylophages from the zone of high fluorine concentration some insect species that prefer newly dead trees move into this zone from the zone of low fluorine concentration, though not to any great extent. Popovichev (1987), during his 9-years' work on control plots, failed to record the presence of the six-teeth bark beetle and the windfall bark beetle in the zone of low fluorine concentration; these beetles seldom occurred in the zone of moderate fluorine concentration but were found more often in the zone of high fluorine concentration.

Species that compose the basis of the xylophagous population in polluted stands and that remain until the forest has decayed are ecologically heterogeneous, though they possess a common trait of resistance to the repellent and toxic effect of fluorine. Grouping insects according to their degree of preadaptation but not to the number of species, a first group can include xylophages, in which the adaptive physiological and ethological behaviour of larvae and imagos corresponds to conditions existing in the stands (and in the phloem and wood of the trees themselves) exposed to fluorine. The large and small pine bark beetles should be ranked first within this group. A second group consists of non-specialized species with a wide ecological amplitude towards variation in the condition of the food plant. They can inhabit trees that are either weakened but still viable or dead but retaining the succulent phloem and the imagos do not feed on the fluorine-saturated bark of branches. This group is represented by the six-teeth bark beetle, the windfall bark beetle, the bright prothorax spruce sawfly and the ridged bronze borer. Some other species could also be included here, e.g. the large and small larch bark beetles, which show, nevertheless, less resistance to a high content of fluorine in the phloem. The conifer pinhole borer occupies a special position since its larvae feed on ambrosia, a fungal mycelium brought into the tunnels by the female, and only partially on wood. Thus, during the 6–7 years of pollution of stands in the zone of high fluorine concentration a population of bark beetles has been formed with a particular composition and showing typical initial accumulation of insect numbers with insignificant involvement of other xylophages. The density of invasion, production and energy of breeding of the main species have reached a relatively high level in this zone (Table 19).

Table 19. The condition of bark beetle populations in pine forests in the zone of (1) high and (2) moderate fluorine concentrations following operation of the smelter for 7 years. (Anisimova and Sokov 1975)

Species	Zone	Nests (Number per dm²)	Young beetles (Number per dm²)	Energy of breeding[a]
Six-teeth bark beetle	1	0.57	4.70	2.42
	2	0.20	3.60	6.86
Large pine beetle	1	1.30	15.00	7.21
	2	1.54	13.90	7.40
Small pine beetle	1	0.37	9.80	14.70
	2	1.77	15.60	5.90
Large larch bark beetle	1	0.96	15.20	3.90
	2	–	–	–
Small larch bark beetle	1	1.23	36.00	14.06
	2	–	–	–

[a] Energy of breeding is determined by multiplication of the coefficient of breeding by a coefficient of insect colonization.

To determine the dynamics of tree invasion by xylophages and tree decline, Anisimova and Sokov used stationary control plots in the zone of high fluorine concentration. The intensity of tree colonization by xylophages and tree decline in the zone of high fluorine concentrations are shown in Table 20. The 7th year was a turning point in acceleration of the invasion. During the next 2–3 years the intensity of xylophage colonization of weakened and declining trees increased somewhat and then stabilized. In this period the density of insect invasion decreased, both from a reduction in species composition and number of different species and from an increasing tendency towards local infestation of the trunks. Furthermore, the coefficient of breeding of principal species in maximally weakened but still-living trees was kept higher than in windfall trees, stumps and stored timber.

At the 10th or 11th year the outbreak changed to a state of chronic suppression or, more exactly, of functional limitation and it remains in this state even now. It is surprising that insects failed to spread into neighbouring weakened but still-viable stands; all species recorded in these regions were of an unusually low intensity of colonization even in irreversibly weakened stands.

Observations by Anisimova and Sokov (1975) showed the following species to be abundant:

In pine: six-teeth bark beetle, small pine beetle, large pine beetle, small larch bark beetle, windfall bark beetle, large larch bark beetle and conifer pinhole borer

In larch: large larch bark beetle, small larch bark beetle, conifer pinhole borer, larch longhorn beetle, bright prothorax spruce sawfly, violet flat sawfly and windfall bark beetle

In spruce: typograph bark beetle, fluffy polygraph, windfall bark beetle, small larch bark beetle, bronze flat sawfly and ridged bronze borer

Subsequently the mass of species was depleted in longhorn beetles, metallic woodborers and some bark beetles. By the time the outbreak had reached a steady state the large and small pine bark beetles and the small larch bark beetle had become the dominant species and the conifer pinhole borer and the windfall bark beetle were also frequently present.

Table 20. The intensity of pine invasion by xylophages and tree decline in the zone of high fluorine concentration

Index (%)[a]	Year following start of smelter operation				
	5th	6th	7th	8th	9th
Annual colonization of trees		2.4	9.7	12.7	14.5
Total tree decline	1.3	5.5	15.2	27.9	42.4

[a] Percent of the total number of trees at the control plot.

Popovichev (1987) examined these forests during 1977–1987 and using permanent experimental plots he failed to find significant variations in the composition of xylophages in the zone of high fluorine concentration during a second decade of the smelter operation. Both species of bark beetles remained dominant, though their number was not large. During this period the numbers of the large pine bark beetle ranged from 2900 to 9700 per ha and of the small pine bark beetle from 3000 to 57800 per ha. Relative spread of the first beetle was 38–100% and of the second beetle 27–88%, expressed as a percentage of the number of trees colonized by xylophages. The conifer pinhole beetle approached close to this value (0–67%), the six-teeth bark beetle was met less frequently (0–33%) and the windfall bark beetle still less frequently (0-26%). In the zone of moderate fluorine concentration the number of bark beetles was much smaller. The numbers of the small pine bark beetles (3-years' mean) exceeded the control by only one-half and of the large pine bark beetle by one-seventh.

On the basis of these observations Popovichev came to the conclusion that, compared to the control, there is an increased number of xylophages in the fluorine-polluted area after a long period of time. Nevertheless, incomplete consumption of the available food supply, the lack of aggressiveness towards weakened stands and the absence of phases characteristic of outbreaks of trunk pests allow this area to be classified not as a chronic outbreak but as a region with a mosaic of xylophages numbers. The mosaic character of this distribution is associated with the condition of stands which, in turn, depends on the distance from the source of emission, the height of land above sea-level and the exposure of slopes.

The conclusion is reasonable. And yet we believe that the situation is best described as an outbreak and not a mosaic distribution of xylophages. It is a new, industrially caused form of outbreak in which one factor, fluorine in the case given, exerts both stimulatory, by providing the source of food, and inhibitory, by being repellent and partially toxic, effects on insect breeding. This double effect of fluorine is the main mechanism regulating not only the number but also the composition of xylophagous species and also the spread of the outbreak within the polluted area.

4.6 Summary

Insects do not play a crucial role in the decline of conifer forests polluted by fluorine-containing emissions. In the zones of high and moderate fluorine concentrations, more than 0.01 mg/m^3, the lethal accumulation of fluorine in conifers is inevitable and varies only in its timing. Therefore, insects can only accelerate the process of drying in irreversibly weakened trees.

Fluorine affects the species composition, ecological groups and numbers of dendrophilous insects. In polluted areas, needle-eating species of forest pests do not breed to produce large numbers and outbreaks of xylophages accelerate the decline of coniferous forests only during the initial years of fluorine pollution. Subsequently, outbreaks of xylophages are characterized by limited species composition, relatively low numbers and low activity of those species capable of inflicting physiological damage on trees. The role of timber pests in forests polluted with fluorine is not large.

Chapter 5

Maximum Permissible Concentrations of Fluorine for Conifers

5.1 Principles of Assessment of Maximum Permissible Concentrations (MPCs) of Toxic Substances for Plants

The assumptions of the limits of acceptable atmospheric pollution proposed by Ryazanov in 1949 have formed the basis for development of public health standards of the content of toxic compounds in the air. MPCs of pollutants in the air for humans is defined as such a level which fails to directly or indirectly affect them. Two standards have been established in the USSR for each pollutant: a single MPC and a daily mean MPC. The single MPC is introduced to prevent reflex responses in humans exposed to a pollutant for a short time (5–20 min) and the daily mean MPC to prevent a repeated absorption effect (toxic, carcinogenic, mutagenic, etc.) of a toxic compound without exact determination of the duration of its impact. Public health standards have been determined for more than 600 substances and for 33 complexes of air pollutants (Nikitin and Novikov 1986).

It does not then seem so difficult to solve the problem of air pollution in human terms. But actually this problem is far more complex because so far humans have only been able to live in an environment which is supported by the plant world and this appears to be extremely vulnerable. The need to establish MPCs of air pollutants for plants has been generally recognized since the 1960s. By that time it had become evident that plants are an indispensable part of man's environment and that concentrations of industrial emissions that can be tolerated by humans are lethal to the majority of plants. Coniferous forests were the first to start dying from air pollution.

Many researchers have contributed to the assessment of MPCs of toxic compounds for plants. Some of them use as a test physiologo-biochemical malfunctions in plants (Nikolaevsky and Miroshnikova 1974; Nikolaevsky 1979; Serebryakova 1980) while others monitor the accumulation of toxic compounds by plants (Reepmeyer and Vetter 1976), a decline in growth (Güde 1981) or root dieback (Ulrich 1980). The studies were performed on whole plants and plant parts using artificial fumigation or by direct measurement in industrially polluted areas. Each of these approaches has its advantages but not all of them meet the basic requirement of MPC as defined here. *The MPC of an air pollutant for plants is the maximum concentration* (*mg*/*m*^{3}) *of pollutant which fails to induce metabolic disorder or cause a decrease in resistance to other injurious factors*. The latter can arise or increase the ecosystem owing to industrial pollution.

As stated by Tarabrin and Korshikov (1986), to assess MPC it is unacceptable to use only a single characteristic as some criteria can result in minor deviations (modification of bioelectric potential, superweak luminescence) when no pathological symptoms are yet present while other criteria can cause profound changes when the pathological process is in progress. Therefore, to provide a more accurate estimation of the general condition of the organism it is necessary to have an integrated index that relates to the state of the whole organism and reflects any alteration in it. Since it is difficult to find such a universal index, a system of parameters is often used for the purpose of providing a complete indication of the onset of pathological changes in the plant organism affected by a pollutant. The values of MPCs for plants, proposed by different researchers, vary greatly in some cases; this is likely to be related to the index chosen, the age and species of plant tested and the impact of additional factors (Table 21).

One of the aims of our research on the phytotoxicity of acid gases was to assess MPC of fluorine for coniferous plants. The work was started in 1981 following a recommendation by the Interdepartmental Committee of the Council of Ministers of the USSR. During these studies we formulated the principles for assessment of MPCs of pollutants for plants and then used them in our work. *To determine MPCs of toxic compounds in the atmosphere it is necessary to use the most sensitive evergreen species of conifers*. Some herbaceous plants and deciduous woody species are also very susceptible to smokes and gases; however, they experience the impact of pollutants on the assimilatory apparatus only during the growth season and most of the accumulated pollutants are removed during autumn leaf-fall. Therefore, toxic compounds affect these plants very little in winter. Furthermore, herbaceous plants may be sensitive to one gas but resistant to another; for instance, gladiolus, tulip and maize can be heavily damaged by hydrogen fluoride but are relatively resistant to sulphur dioxide while chicory, tomato and lucerne are highly sensitive to sulphur dioxide and show low sensitivity to hydrogen fluoride (Zimmerman and Hitchcock 1956; Ten Houten 1974).

Deciduous trees are less suitable as bioindicators of air pollution not only because of their higher resistance to pollutants but also because of the lack of correlation between the accumulation of, for instance, fluorine in the leaves and the degree of tree weakening. Conifers (at least boreal species), on the other hand, show a distinct correlation between these parameters. To determine MPC the most common coniferous species are usually used. In Europe these are spruce and sometimes fir and under the conditions of

Table 21. MPCs (mg/m^3) of toxic atmospheric compounds for plants, recommended by different countries

Compound	Period of impact		Reference
	Short-term	Long-term	
Sulphur dioxide	0.300	0.015	Nikolaevsky (1984)
	0.020	–	Nikolaevsky and Miroshnikova (1974)
	0.080	0.020	Serebryakova (1980)
	1.300 (3 h)	–	JAPCA (1971)
	0.260 (24 h)	0.060 (average annual)	JAPCA (1971)
	0.400	0.140	TA Luft (1974) (after Guderian 1979)
	–	0.052 (average monthly)	IUFRO (1974)
	–	0.024 (average annual)	IUFRO (1974)
	–	0.006 (average per century)	IUFRO (1974)
	0.150 (24 h)	–	Materna (1975)
	0.500 (30 min)	–	Mielke (1977)
	0.150 (24 h)	–	Mielke (1977)
	0.075 (30 min)	0.025 (average annual)	Latocha (1983)
	0.050 (24 h)	–	Latocha (1983)
Sulphuric acid vapour	0.100	–	Nikolaevsky and Pershina (1981)
	0.100	0.030	Nikolaevsky (1984)
Hydrogen fluoride	0.020	0.003	Nikolaevsky (1984)
	0.004	0.002	TA Luft (1974) (after Guderian 1979)
	0.00155 (1 week)	0.0005 (growth period)	Weinstein (1971) (after Guderian 1979)
	0.00270 (24 h)	0.00078 (1 month)	Weinstein (1971) (after Guderian 1979)
	0.00350 (12 h)	–	Weinstein (1971) (after Guderian 1979)
	0.020 (30 min)	–	Mielke (1977)
	0.005 (24 h)	–	Mielke (1977)
Soluble solid fluorides	0.030 (30 min)	–	Mielke (1977)
	0.010 (24 h)	–	Mielke (1977)

Chlorine	0.006 (conifers)	0.015 (conifers)	Serebryakova (1980)
	0.080 (deciduous)	0.020 (deciduous)	Serebryakova (1980)
	0.025	–	Nikolaevsky and Pershina (1981)
	0.025	0.015	Nikolaevsky (1984)
	0.030 (24 h)	–	Mielke (1977)
	0.100 (30 min)	–	Mielke (1977)
Hydrogen chloride	0.200	0.100	TA Luft (1974) (after Guderian 1979)
Hydrochloric acid vapour	0.050 (30 min)	–	Mielke (1977)
	0.015 (24 h)	–	Mielke (1977)
Nitrogen oxides	0.050	–	Nikolaevsky and Miroshnikova (1974)
	0.020	–	Nikolaevsky and Pershina (1981)
	–	0.030 (conifers)	Serebryakova (1980)
	–	0.050 (deciduous)	Serebryakova (1980)
	0.040	0.020	Nikolaevsky (1984)
	0.100 (30 min)	–	Mielke (1977)
	0.040 (24 h)	–	Mielke (1977)
Hydrogen sulphide	0.020	–	Nikolaevsky and Pershina (1981)
Carbon monoxide	4000	–	Nikolaersky and Pershina (1981)
	5.000	3.000	Nikolaevsky (1984)

a harsh continental climate common pine is used. In pine the damage proceeds gradually, especially in the zones of high and moderate fluorine concentration; its susceptibility to pollutants is similar to that of lichens (Guderian 1979; Freedman and Hutchinson 1980). To monitor the degree of damage in spruce is more difficult as the symptoms appear irregularly in the crown.

MPC should not cause significant differences in the functional and structural parameters measured between fumigated and control plants during the period of high physiological activity. To measure MPC the most sensitive indicators of the trees' condition should be monitored. MPC of a pollutant is determined in whole plants by artificial fumigation in field chambers with 10- to 20-year-old conifers, control assays and measurements on young and mature trees in areas polluted to varying degrees by industrial emissions. Isolated parts of plants should not be used for several reasons. First, pollutants affect all parts and organs of the plant to a varing extent; they are transported by downward and upward flow within the plant and can be accumulated in certain tissues or removed and released the environment. The process of physiological weakening happens slowly and changes observed in the needles in the initial stages, for instance, may not correspond to the general condition of the plant. Second, in addition to a direct impact, pollutants also affect the plant indirectly, synecologically and this effect is very important. For example, when a pollutant is accumulated in the soil, especially in the vicinity of the source of pollution emission, it may affect the plant directly and also indirectly by changing the chemical composition and structure of the soil, causing the death of soil microorganisms and decreasing the volume of mycorrhizae and thin roots.

The response of the tree to industrial pollution is ultimately the result of a whole complex of ecological factors that are enhanced or declined by the effect of toxic compounds. Damage may be enhanced by high solar radiation, elevated temperatures and air moisture, fogs, lack of winds, prolonged temperature inversion and anomalous climatic conditions such as spring frosts or frequent drought periods. *The established MPC of a pollutant should not, in combination with unfavourable ecological factors, bring the plant to a condition of "unsteady balance" and thus allow attack by pests and pathogenic microorganisms.*

To determine MPC of a pollutant with a short-term impact the same principles are applied, though most attention should be paid in this case to leaf diagnosis by determining needle damage and pollutant accumulation in the needles. The needles experience the greatest effect of a short-term impact of an increased pollutant concentration. The period of the short-term effect in plants depends on the duration of emissions during occasional smelter failure and usually does not last more than a few days.

The MPCs of pollutants for plants are determined by plant genetic and ecologo-geographical resistance to adverse factors in general. Ecological conditions, such as climate, affect each species by ranging from the most favourable for growth to unsuitable. Consequently, variations both in forest injury and individual tree resistance to damage are very large over the vast territory of the USSR. For instance, in the southern taiga larch can tolerate complete defoliation by insects for 3 consecutive years, in the middle taiga not all larch trees can tolerate a twofold increase in needle loss and in Yakutiya even a single defoliation can lead to larch trees drying in the following year.

In our opinion two kinds of standards, *general* and *regional*, should be established for MPCs of pollutants in the case of woody plants, as assessed in conifers. The general standard is the principal one, assessed in regions of favourable climatic and soil conditions for coniferous forests. The regional standard would be established for particular industrial centres to consider the particular conditions of the area and indicate the variation from the general standard. In many regions forests are exposed to the simultaneous impact of many toxic compounds that are emitted into the atmosphere. The number of such regions is inevitably increasing. In large industrial centres it is necessary to perform special studies on the combined action of pollutants which usually inflict more severe damage on plants by synergistic or additive effects.

5.2 The System of Criteria for Assessment of Levels of Pollutant MPCs in Woody Plants

Experience over many years has shown that to determine MPC of a pollutant for coniferous plants it is necessary to use not one criterion, even if it seems to be the most reliable factor, but a combination of characteristics of the plants' condition, including visual (needle damage, morphometric changes to needles and auxiblasts and formation of additional shoots from dormant buds) and physiologo-biochemical parameters as well as direct accumulation of the pollutant in the needles. Morphometric parameters, being mostly integral in their nature, reflect physiologo-biochemical variations and therefore can serve as indicators of a critical point in plant development. The most indicative and easily determined physiologo-biochemical parameters in trees exposed to low hydrogen fluoride concentrations appear to be changes in the activity of peroxidase, in the intensity of respiration and in the content of photosynthetic pigments and ascorbic acid. No doubt they could also apply to exposure of trees to other acid pollutants.

Only healthy trees are suitable for artificial fumigation in chambers. Care must be taken not to use subjects that are even slightly infested by pathogens as this can lead to incorrect results. Plants that have been exposed to some diseases may show a possible increase in resistance to pollutants (Ten Houten 1974) whereas colonization by insects indicates, as a rule, a decrease in resistance and a weakening of the organism.

No well-expressed common changes in physiologo-biochemical and morphometric parameters have been found with exposure to different acid gases; however, there are certain similarities in the character of needle damage, in the duration of the latent effect of low detrimental doses and in the accumulation and removal of the pollutant from the plant.

Morpho-physiological characteristics of the plant vary during vegetative growth, and during the process of ontogenesis as a whole, and consequently the level of gas sensitivity of separate organs changes; there are relatively insensitive periods alternated by critically sensitive periods (Kulagin 1985). According to Wentzel (1963) increased sensitivity is observed in the final stage of growth in young conifers during the period of intense growth and it remains until the early stage of tree maturity. It is known that the transition from winter to the season of vegetative growth is the most sensitive period for conifers of all ages. Young growing needles on current-year shoots are highly sensitive to pollutants. Nevertheless, we believe that the most suitable time to conduct diagnostic physiologo-biochemical studies, bearing in mind the stability and intensity of the processes mentioned above, is in the first half of July. At this time we also measured the content of fluorine in the needles to assess its maximum permissible accumulation. The morphometric analysis of needles and shoots should not be done until the end of July.

It should be borne in mind that determination of the direct accumulation of fluorine in the needles (for diagnostic purposes) is specific. Thus, fluorine, unlike sulphur and chlorine, does not belong to the biogenic elements and its content in plants is extremely low under normal conditions. Therefore, an increased concentration of this pollutant in tissues, as compared to control tissues, may be a reliable index of the impact of fluorine emissions as a cause of tree decline. In the case of sulphur and chlorine, chlorosis and even necrosis of plant leaves may be observed because of extra accumulation of these elements in amounts only slightly exceeding the average normal level; this could be accounted for by extreme variation in their content under normal conditions (Guderian 1979).

The complexity in assessing MPCs of substances that exert a cumulative effect on plants should be emphasized again. As already stated the cumulative effect is possible under long-term exposure of plants to low doses of fluorine in zones of low levels of emission. At the same time, the plant can detoxify a certain amount of fluorine and tolerate it without any adverse effects to its performance. Therefore, it is essential to determine the level of fluorine which would be lower than the concentration inducing a cumulative effect.

5.3 MPCs of Fluorine for Conifers

Our aim was to determine the experimental MPC and the natural MPC (in stands polluted by fluorine-containing emissions) of fluorine for the principal forest-forming coniferous species of the Baikal area; these species are Siberian and Dahurian larches, common pine and Siberian spruce. With this aim in view, a set of experiments was carried out by artificially fumigating trees in field chambers with a controlled regime of pollution. In spring, 10- to 20-year-old trees were planted in the chambers and fumigated during the next growing season only. The dynamics of fluorine accumulation in tree needles in terms of fluorine concentration in the atmosphere were concurrently examined in premature and mature stands exposed to emissions from aluminium smelters. The trees in chambers were fumigated with hydrogen fluoride for 18–20 h/day during June–August for 3 years. To control the concentration of fluorine in the chambers daily samples of air were assayed.

The following principal characteristics of trees exposed to hydrogen fluoride were monitored: the lack of or appearance of chlorosis and necrosis in the needles, needle length, the content of fluorine in the needles, bursting of dormant buds and development of substitutive auxiblasts (the symptom of reduced performance of the assimilatory apparatus), linear increment of shoots and physiologo-biochemical changes in the needles. Peroxidase activity, respiration intensity and content of photosynthetic pigments and ascorbic acid were the first physiologo-biochemical parameters to be examined. The content of fluorine in the needles was measured in a photoelectrocolorimeter following preliminary distillation of the ashed sample. Physiologo-biochemical parameters were assayed by conventional methods.

The first stage of our experiment was aimed at detecting (by the series of gas chamber experiments) the maximum concentration of hydrogen fluoride in the air under which the trees did not show visual symptoms of damage over three growing seasons. It was found to be 0.001 mg/m^3. This dose induced accumulation of fluorine to 4–7 mg% dry weight in the needles. No accumulation of fluorine was found in the soil under these circumstances and hence, no additional fluorine effect was exerted on the trees.

The next step was to verify the MPC of hydrogen fluoride using the findings of physiologo-biochemical analysis of plants from two growing seasons. The parameters of the tree state that were measured are the first to alter under the influence of fluorine. Physiologo-biochemical processes in plants have been shown to exhibit no or very little change when the fluorine concentration in the air is not more than 0.0005 mg/m^3 and when fluorine is accumulated in the needles of larch up to 4 mg% dry weight and in pine and spruce up to 6 mg% dry weight. Under such concentrations no significant changes were found in the activity of peroxidase, intensity of respiration or

Table 22. Physiological-biochemical parameters of needles exposed to artificial fumigation with 0.0002–0.0005 mg/m^3 hydrogen fluoride

Species	Activity of peroxidase (arbitrary units)	Intensity of respiration (mg CO_2/g FW/h)	Pigments (mg/g DW)	Ascorbic acid (mg% FW)
Siberian larch				
Control	0.28 ± 0.03	1.41 ± 0.05	2.40 ± 0.04	57.05 ± 2.29
Experiment	0.30 ± 0.03	1.43 ± 0.04	2.49 ± 0.07	58.73 ± 1.40
Dahurian larch				
Control	0.52 ± 0.04	1.47 ± 0.05	2.12 ± 0.04	41.16 ± 1.76
Experiment	0.57 ± 0.02	1.52 ± 0.04	2.25 ± 0.04	46.35 ± 1.25
Common pine				
Control	0.99 ± 0.04	1.14 ± 0.03	2.33 ± 0.05	33.48 ± 1.34
Experiment	1.07 ± 0.03	1.15 ± 0.03	2.35 ± 0.06	33.15 ± 1.45
Siberian spruce				
Control	0.26 ± 0.03	0.71 ± 0.04	3.72 ± 0.06	34.53 ± 1.49
Experiment	0.31 ± 0.03	0.76 ± 0.04	3.61 ± 0.14	36.60 ± 1.01

the content of pigments and ascorbic acid (Table 22). Presumably plants can detoxify such an amount of fluorine without suffering any metabolic disorder. Furthermore, the plants did not lose the ability to resist attack by xylophagous insects or infection by pathogenic microorganisms.

The level of permissible short-term concentration of fluorine in the air was determined in field chamber experiments by monitoring the accumulation of fluorine in the needles. It was found that the permissible content of fluorine of 4–6 mg% dry weight, assessed in long-term experiments, was accumulated by needles exposed to 0.004 mg/m^3 hydrogen fluoride for 6–7 days.

The experimental level of permissible concentration of fluorine in the needles was compared to the content of fluorine in plants growing in a zone of long-term impact of very low fluorine emissions. The results illustrated that under such conditions the content of fluorine in the needles was not more than 4 mg% dry weight in larch and 6 mg% dry weight in pine and spruce and that shoot growth did not show any significant decrease in linear increment; increasing the fluorine concentration in the needles of pine and spruce to 8–9 mg% dry weight and of larch to 7–8 mg% dry weight resulted in a statistically significant drop in shoot growth.

In conclusion, the following maximum permissible atmospheric concentrations of fluorine have been recommended for conifers: 0.0005 mg/m^3 for long-term exposure and 0.004 mg/m^3 for short-term exposure, i.e. not more than 1 week per year. MPC of fluorine in the needles of larch is 4 mg% dry weight and in pine and spruce is 6 mg% dry weight.

Chapter 6

The Use of Conifers for Planting Protection Zones

6.1 Some Aspects of Designing the Planting Protection Zone

The present-day expansion of utilization of natural resources makes it necessary to coordinate the increase in industry with its effects on ecosystems. In connection with this, the principal tasks are, first, to heal the wounds suffered by nature by means of forest culture and natural reafforestation processes and, second, to enhance the ability of woody plants and forest communities to accumulate pollutants and detoxify the atmosphere (Kulagin 1980; Liebold and Ranft 1981). This can be done by selecting individual gas-resistant trees, establishing plantations of gas-resistant clones and choosing relatively gas- and smoke-resistant ecological populations. The latter method is the most popular as it saves time and reduces the expense of producing plant material that is resistant to industrial pollution.

The expansion of industrial capacity and increased concentration of industries, on the one hand, and the low level of improvement in waste purification systems, on the other hand, have resulted not only in total air pollution but also in the expansion of areas polluted by particular industrial enterprises. As a consequence, the concentration of pollutants has risen in inhabited areas situated far from the source of emission and degraded areas of forest have expanded. Under these conditions planting protection zones in areas between industrial enterprises and populated regions can no longer be efficient; moreover, the present designation of such zones should not be limited only to protection of the human population against harmful emissions. The protection zones around industrial plants that emit toxic compounds into the atmosphere are intended to decrease the adverse impact of these compounds not only on people but also on the surrounding environment.

A protection zone which essentially consists of arboreal and shrub plants is designed simultaneously with the industrial enterprise and forms an integral part of it. Hence, the industrial plant could not be put into operation prior to the completion of planting the protection zone. Occasionally design and formation of the protection zone has occurred after the industrial plant has begun operation; this resulted in a significant rise in costs and a lowered protection ability of the plants, especially if the existing stands had already started to dry out. It is obviously not possible to restore declined and irreversibly weakened forests under conditions of continuous pollution of the environment.

Long-term observation of stands polluted by fluorine-containing emissions from aluminium plants allow us to propose some suggestions on the design of protection zones in general and on the conditions determining their parameters and peculiarities in the formation of landscapes. The protection zone consists of three parts: industrial, populated and intermediate. The industrial part includes the territory of the plant and its surroundings, the size and configuration of which are estimated by some parameters; the populated part embraces the populated territory and surrounding forest-park belt; and the intermediate part is situated in between the plant and the populated territory and extends for 3–8 km from the plant. The belts of trees along highways and railways radiating from the industrial plant to as far away as 20 km could be included in the protection zone.

The size and configuration of the entire protection zone and its parts are determined by: (1) the composition and amount of emitted pollutants and their toxicity to the most sensitive species of plants (with regard to the effects of addition, synergism and domination of the components); (2) the relief of the locality, the density of the river network and availability of large reservoirs; (3) recurrent winds and their velocities and the duration of calms; and (4) afforestation of the area, species composition and the age of stands. These parameters can be obtained by preliminary studies and calculations and serve as a basis for the proposal and feasibility study of protection zones.

During the design of industrial centres the combination of hydrogen fluoride with chlorine or sulphur dioxide in emissions should be excluded since it could lead to a strong synergistic effect. This effect could also be brought about by the impact of chlorine and sulphur dioxide mixtures. The combination of all three pollutants is still more hazardous since it is responsible for several superimposed synergistic effects. When fluorine acts in combination with nitrogen oxides or carbon monoxide the calculation of pollutant toxicity should be done in terms of fluorine as it exerts the dominant effect in this case. The production of pure silicon, which usually accompanies aluminium production, results in silicon powder pollution; this should be removed from the area polluted with fluorine and beyond as it is harmful by itself and also contributes to the formation of toxic fogs (minute particles of silicon serve as nuclei of vapour condensation), thus expanding the zone of chemical burn of plants.

The topography of the locality and winds are basic factors responsible for the dispersion of gaseous and dust-based fluorine compounds over certain distances and therefore for the size and configuration of the polluted area and zones of high, moderate and low fluorine concentrations within the area. Design of the construction and disposition of a large-scale aluminium plant of 20 or more buildings in mountainous taiga and mountainous forest steppe regions of Eastern Siberia, even with provision for perfect purification of emissions, should take into account that the closest populated location should be at least 15 km from the plant, at a lower altitude than the plant, separated by a watershed and have good aeration conditions.

One of the faults in plant design, concerning the distance from populated locations, was the reference only to average annual frequency of total wind and its velocity. No account was taken of the specific character of local air circulation (sometimes fairly complex), dynamics of the seasonal damage to woody plants and the fact that infrequent and weak air flows of the same direction are no less dangerous than prevailing winds, since they may contain a large amount of fluorine and transport it at a high concentration over significant distances. Calms are the most dangerous for stands that neighbour aluminium smelters.

In designing the landscape formation on broken ground, account must be taken of the peculiarities of hydrogen fluoride distribution over the polluted area. Hydrogen fluoride inflicts damage on coniferous stands mainly on windward slopes of open uplands. Hydrogen fluoride shows poor or no stagnation in lowlands and closed valleys (troughs), while in the presence of other heavier pollutants, hydrogen fluoride is inclined to stagnate in lowlands. This fact is taken into account when considering the position of shielding and filtering plantations. It is also very important to make allowance for a cumulative effect of fluorine and to provide favourable soil conditions for all artificial plantations, especially in the zone of high concentration of fluorine, to ensure adequate cultivation.

6.2 The Feasibility of Using Coniferous Species in Protection Zones. The Dahurian Larch Form of Higher Fluorine Resistance

A search for forms of coniferous trees that are resistant to industrial pollution is a necessary task in terms of enhancing the resistance of forests to gases. Conifers absorb considerable amounts of pollutants. Most of them have needles from many years of growth and are capable of performing filtering functions for a whole year. Moreover, as numerous papers report, the state of coniferous species in industrially developed countries is of great concern since conifers are extremely sensitive to pollutants. Many papers report an urgent need to restore native forests with prevailing coniferous species (Greshta 1970; Kawecka 1973; Kluczyński 1975; Materna 1975, 1978; Getko et al. 1978; Klock 1980; Wentzel 1980, etc.).

The data given in the second and third chapters indicate that conifers cannot tolerate high concentrations of fluorine in the atmosphere. The representatives of the genera studied differ in their general resistance to fluorine and in their susceptibility to chemical burn. Larch, which possesses undeniably superior resistance to pollutants and regenerating ability, is less resistant to needle burn and to raising green shoots and therefore in young trees it is inferior to pine (and sometimes to spruce) in its total resistance to the effect of fluorine. Mature and old pine trees usually decline earlier than larch, while its young shoots are better protected from chemical burn than in other conifers. However, on the whole, all conifers, with the exception of particular relatively resistant forms that result from growing under conditions of naturally elevated toxicity of the atmosphere and soil, are not suiable for use in plantations in the industrial part of the protection zone and are of little use in the intermediate part. Nevertheless, owing to a rather wide range of individual resistances in conifers, one can obtain more resistant planting material by selecting trees in nurseries. In the zone of moderate fluorine concentration they can be used as underbrush and as the lower stratum in multi-stratum forest belts (in internal rows), as well as in shielding and filtering artificial plantations on windward edges of naturally growing coniferous forests.

In the case of coniferous afforestation of large areas in the zone of moderate fluorine concentration, the cultures of selected plants should be alternated by bands, three to four rows deep, of large deciduous trees with underbrush. The bands should be arranged 40–100 m apart, depending on the distance from the industrial plant, exposure of the slope and direction of prevailing winds. It is advisable to alternate coniferous plantations with taller shrubs which could subsequently form underbrush. The planting of protective bands of large deciduous trees is also necessary when highly resistant forms of conifers are used. Outside of the zone of lethal accumulation of fluorine by trees but still within the zone of pollution, especially in the forest-park belt around populated locations, coniferous trees should form the basis of plantations created.

In the mid 1970s we started a search for conifers that possess enhanced resistance to fluorine in regions under the constant influence of an excess of fluorine anions. First, we considered the use of trees growing close to active vulcanoes in Kamchatka; however, we could not manage to get planting material from this region. Then the idea arose to test fluorine resistance of conifers growing at outcrops of fluorite deposits. In 1977, 12- to 15-year-old Dahurian larch trees were delivered from one of the fluorite mines in the Chita region; these trees were exposed to fumigation for 4 years in field chambers at the laboratory field station. It appeared that trees from this population (hereafter called the fluorite population) had significantly higher resistance to fluorine than did the control trees, which were Siberian larch from the Baikal area (Tunka valley) and Dahurian larch from the Transbaikal region. The experiments demonstrated that Dahurian larch was generally of somewhat higher resistance to hydrogen fluoride than Siberian larch.

The trees were fumigated with hydrogen fluoride in the chambers for 18–20 h/day during the whole of the growing season and the concentrations of gas used were 0.10 and 0.35 mg/m^3: Air samples were collected from the chambers at 8 a.m. each day to monitor the concentration of fluorine. The toxic effect of hydrogen fluoride inflicted damage on the needles of control plants of Siberian and Dahurian larch by 10 and 20–40% after 8 and 24–34 days of fumigation respectively, while the trees of the fluorite population in the same chambers suffered less damage: only 5% of needles were injured after 34 days of fumigation. By the end of the growing season the maximal damage of the crown was as low as 15%. Identical results were obtained in fumigation experiments in the second and third years.

The average content of fluoride ions in the soil of the fluorite mine was about 0.07%. Analysis before fumigation revealed that under natural conditions of growth the fluorine concentration in the needles of the fluorite population of Dahurian larch was almost three times higher than in the common form of this species from the Transbaikal region, while the intensity of fluorine accumulation during the process of fumigation was almost two times lower than in the common form (Table 23). In the subsequent 2 years of fumigation Dahurian larch of the fluorite population showed either only slight necrosis of needles or almost complete absence of any visible symptoms of damage.

The study of physiologo-biochemical responses in trees of the fluorite population of Dahurian larch and in control trees (Table 24) showed that there were less changes to a number of metabolic processes in the fluorite population trees following exposure to fluorine than there were in the control trees; this suggests higher resistance of the fluorite population. Some increase in the content of chlorophyll and an essential lack of change in the amount of reducing sugars implies that the photosynthetic apparatus of the fluorite population is less vulnerable to the effect of fluorine. The intensity of oxidative processes, monitored by peroxidase activity, also did not rise significantly in the fumigated trees of the fluorite population. Less difference was observed in changes of respiration intensity in the needles and in the

Table 23. Accumulation of fluorine and degree of needle damage in trees exposed to fumigation with 0.10 mg/m^3 hydrogen fluoride for 35 days

Species	Fluorine content in needle (mg% DW)		Needle damage (%)
	Control	Experiment	
Siberian larch (Tunka valley)	2.17 ± 0.115	14.05 ± 0.200	30–40
Dahurian larch (common form)	2.30 ± 0.089	15.25 ± 0.118	20–30
Dahurian larch (fluorite form)	6.62 ± 0.130	19.12 ± 0.158	Not more than 5

Table 24. Variation in some physiological-biochemical parameters between different forms of Dahurian larch exposed to fumigation with 0.10 mg/m^3 hydrogen fluoride for 35 days

Parameter	Common form		Fluorite form	
	Control	Experiment	Control	Experiment
Peroxidase activity (arbitrary units)	0.25 ± 0.015	0.83 ± 0.014	0.63 ± 0.011	0.82 ± 0.015
Respiration intensity (mg CO_2/g FW/h)	1.78 ± 0.027	1.87 ± 0.033	1.89 ± 0.034	2.00 ± 0.029
Chlorophyll content (mg/g DW)	2.04 ± 0.029	1.44 ± 0.022	2.18 ± 0.028	2.77 ± 0.030
Water content (% FW)	65.43 ± 0.38	59.52 ± 0.45	61.10 ± 0.47	59.53 ± 0.44
General acidity (% DW)	2.14 ± 0.022	2.23 ± 0.021	1.76 ± 0.019	1.61 ± 0.015
Ascorbic acid content (mg% FW)	56.35 ± 0.48	30.52 ± 0.62	35.90 ± 0.35	26.12 ± 0.37
Reducing carbohydrate content (% DW)	2.80 ± 0.017	2.20 ± 0.031	2.47 ± 0.025	2.59 ± 0.014
Total nitrogen content (% DW)	2.15 ± 0.023	4.15 ± 0.032	2.67 ± 0.026	2.77 ± 0.043
Resins				
Total content (% DW)	6.7 ± 0.08	2.9 ± 0.009	5.6 ± 0.18	8.7 ± 0.12
Fraction of neutral substances (% DW)	3.9 ± 0.11	1.5 ± 0.07	2.9 ± 0.10	7.2 ± 0.13

content of ascorbic acid between the different larch forms but rather drastic changes were observed in the intensity of resin formation. In the fluorite population of larch the total content of resins rose by an increase in the fraction of neutral substances, whereas acidic resin amounted to as little as 1.5% (2.7% in the control). The transformation of resin synthesis leading to an accumulation of neutral substances indicates the existence of protective behaviour and even suggests a certain rise in tree resistance to possible attack by pests. In contrast, trees of the common form of larch showed a diminished content of resins in general and of the neutral fraction in needles when exposed to the same concentration of hydrogen fluoride.

These differences in physiologo-biochemical transformations in the two forms of Dahurian larch are related to the different degree of their damage. As the needles of the resistant form were a little necrotic, which can actually imply the initial stage of damage, there was a certain acceleration of metabolism in this form. The changes in physiologo-biochemical parameters induced by 30% damage to the needles of the common form already indicate metabolic disorder. The mechanisms responsible for the higher resistance to fluorine of trees of the fluorite population of Dahurian larch are obscure. There is an opinion that the differences are associated with the different rate of toxic gas absorption (Nikolaevsky 1979). In our experiments the differences in the rate of fluorine accumulation between common and fluorite forms of larch are large (see Table 23); however, no significant differences have been found in the intensity of gas exchange (see Table 24). Another argument, given by Guderian (1979), is that the absorption of hydrogen fluoride and hydrogen chloride (in contrast to sulphur dioxide absorption) fails to correlate with the activity of gas exchange and proceeds much more rapidly. It is assumed to be related to the ability of fluoride ions to penetrate into the leaf directly through the cuticle and epidermis. The fluorite population of Dahurian larch had been adapting not to the effect of atmospheric fluorine but to the effect of soil fluorine for many millenia. It can be presumed, for the present, that during natural selection this population had developed a more intensive efflux of fluorine into the needles prior to their autumn fall.

The trees of the fluorite population are rather undersized in areas of their indigenous growth. The radial increment of the trunk is also diminished and there are early disturbances to the ratio of the lengths of basic to additional branches; the needles are shortened and have thicker cuticles and more thickened epidermal and hypodermal cell walls (especially in auxiblasts) than in the common form of Dahurian larch. The hypodermal sclerenchyma is developed as a median belt on the upper side of needles and along highly protruding carinae on the lower side.

The described peculiarities of the fluorite population, except for anatomical structure of the needles, are not genetically inherited and result from the constant impact of toxins in the environment. When the plants are transported from the zone of enhanced soil fluorine content into normal conditions linear and radial growth are readily restored to values typical for the species. In our experiments, during the 7 years after the 3-year fumigation of trees of the fluorite population with hydrogen fluoride a 4- to 5.3-fold increase in height was observed (annual 36–82 cm increment of the central stem) as well as a 5-fold increase in the radial trunk increment at 60 cm height. The needles resumed their normal size, reaching 30 mm in auxiblasts and 42 mm in brachyblasts by the end of July; however, the walls of epidermal and hypodermal cells remained thickened.

In 1980, after testing trees of the fluorite population of Dahurian larch in chambers, some of them were planted in the zone of high pollution of the aluminium smelter. No signs of damage were observed in these trees during all subsequent years. They were characterized by high increment and normal length of the needles.

6.3 Selection of Coniferous Nursery Trees

Owing to genetic heterogeneity of populations, the specimens composing them possess very unequal resistance to injurious factors. As already mentioned, individual differences in conifer resistance to fluorine are very great.

The period of decline of trees of the same species and of the same age growing in the same site in the zone of high concentration may vary by several years and in the zone of moderate concentration single trees of Siberian larch can retain their viability for 10 years following the onset of decline in other trees of the same age. Mamaev and Makarov (1976), examining individual variability in common pine by experimental fumigation with sulphur dioxide, found that about 6–7% of trees were highly gas resistant, 30–35% moderately resistant and 60% had low resistance. Considerable variation in the damage by pollutants (SO_2, O_3 and HF) has also been found in some clones of eastern white pine (*Pinus strobus*) (Berry 1973). The clones were found to be sensitive to all gases, resistant to all gases or intermediate. Furthermore, highly sensitive trees were used as bioindicators and the most resistant trees as sources of material for establishing pollutant-resistant seed plantations of eastern white pine. Similar work has been performed by Karolewski and Bialobok (1979) with clones of common larch. Shoots that had been cut from 20 different larch clones were fumigated with sulphur dioxide, ozone, hydrogen fluoride and mixtures of these gases. The results showed that some clones exhibited enhanced resistance to the effect of two or even three of the gases. In northern Britain an attempt was made to discover gas-resistant species of trees, including conifers, by raising nursery trees from seeds of different origin in an industrial area (Lines 1981). The examples cited illustrate that to determine the gas resistance of species and individuals, researchers prefer methods of monitoring the direct effect of pollutants on plants followed by a further control over the plant state. Available methods of indirect estimation of gas resistance are not so reliable.

Specimens that are found to be resistant can be further propagated vegetatively, for instance by grafting on seedlings of cuttings from gas-resistant specimens (Tzschacksch 1983; Weidenbach 1984). Successful results in developing methods of selection of resistant woody specimens have been obtained in Germany (the Tarandt term) where extensive measures are taken to preserve and restore coniferous and deciduous stands in polluted areas (Ewert et al. 1981; Liebold and Ranft 1981; Tzschacksch 1983). The system of selection of resistant planting material for zones exposed to the impact of sulphur dioxide is most widely used. Using transportable gas chambers nursery trees were chosen directly from rows in forest nurseries; in particular 4- to 6-year-old spruce trees of increased resistance to this gas were selected (Ewert et al. 1981).

To estimate individual resistance of conifers to the effect of fluorine we used two methods in our experiments: prolonged fumigation with low concentrations of hydrogen fluoride in chambers; and application of small doses of sodium fluoride into the soil (the root effect). In the case of the effect via leaves, 3- to 5-year-old pine trees were fumigated in polythene chambers where 0.1-0.3 mg/m^3 fluorine was applied each day for 15–20 min during June–August. The experiments were performed for 3 consecutive years. The degree of gas resistance was estimated by the extent of crown damage in terms of percentage of necrotic needles. After the first 3 months of fumigation no visible damage was observed; in the second year some specimens exhibited needle damage as a slight necrosis of the apical part; and finally in the third year a half of the trees had from 5 to 50% crown damage. A portion of the trees without any damage were transplanted to the vicinity of the smelter to test for their resistance to fluorides under conditions of industrial air pollution. One year after transplantation no necrotic needles were found in these trees though the needles of the current-year shoots were two to three times shorter than the normal needles.

The second method of identifying fluorine-resistant trees involved testing the effect of fluoride ions present in the soil in excess. To do this, 10- to 12-year-old pine and spruce trees were watered with solutions of sodium fluoride of increasing concentration, from 0.01% to 0.10% by the end of the experiment. The trees were watered at intervals of 4–5 days from June to August (20 times in total) over 3 years. The volume of solution for the single treatment of a single tree was 2 litres. The first 3-month treatment did not inflict any visible damage on trees until the end of the growing season. By the next spring, 25% of pine and spruce needles showed a little damage, apparent as slight and partial necroses of the ends of single needles. The second 3-month treatment was followed by an increase in visual damage by the end of the growing season and 60–70% of trees had necrotic needles (up to 40–50% of the crown). By the onset of treatment in the third year of the experiment some trees had begun to decline, while 15% of pine and 25% of spruce still showed no visible damage (Table 25). Some of the resistant specimens were transplanted into the zone of high fluorine concentration close to the aluminium smelter. No damage of the needles was observed in these trees over the next 2 years. Unfortunately the experimental plot was then destroyed.

The use of gas chambers for choosing specimens of woody plants resistant to pollutants is an expensive and inflexible method. As to the use of a study of long-term root effects, we consider this method to have

Table 25. Species differences in conifer damage

Species	Number of seasons with treatment[a]	Trees (%)		
		Undamaged	Damaged	Dried
Common pine	1	75	25	–
	2	25	70	5
	3	15	75	10
Siberian spruce	1	75	25	–
	2	35	60	5
	3	25	60	15

[a] Each season with treatment lasted from June to August.

advantages over a single treatment with a high concentration of fluorine. The conifers are known to tolerate more readily a short-term impact of high concentrations of gases than they are a long-term exposure to low concentrations. Therefore, long-term testing of trees with low concentrations of fluorine can reliably detect plants of higher resistance and simultaneously harden them to the toxic effect of the pollutant. This method may be applied to obtain relatively gas-resistant planting material in forest nurseries.

6.4 Deciduous Arboreal and Shrubby Plants Resistant to Fluorine

In addition to coniferous species, a test for resistance to fluorine has been performed with local and introduced deciduous trees and shrubs. As a result, 24 species of plants have been identified which are resistant to exposure to fluorine or possess a high reducing ability.

To create protective green zones around large-scale industrial centres is hard work under the conditions found in Eastern Siberia. The recommended species must have not only high gas resistance but also frost resistance and be good at reproducing. Therefore, plants characterized by an ability to survive under the climatic conditions of the region were chosen for testing. To guarantee success the plants were tested under more stringent conditions (constant high concentration of fluorine, high air moisture and enhanced radiation) than those they were intended for. The tests were performed by artificial fumigation in large field chambers for 3 months in the summer. The plants transplanted into open chambers were allowed to root for 1 year, then, at the end of May, the chambers were closed and filled with the gas, the concentration of which was gradually increased over 3 weeks up to $0.3\,mg/m^3$. The gas resistance was determined by the degree of leaf damage (percentage of necrosis). Plants with not more than 10% necrosis of the total leaf area were considered as slightly injured, with 40% as moderately injured and with 40–100% as severely injured. The gas-accumulating ability of plants was estimated by fluorine accumulation by the leaves at the end of the fumigation period and the behaviour of plants was analysed after the winter.

The plants recommended for planting in protection zones (Table 26) differ in their degree of damage, amount of fluorine absorption and reducing ability. The species with a low ability for fluorine absorption and that show a post-burn decrease in winter hardiness are not included in Table 26.

Analysis of the content of fluorine in leaves showed extremely wide variation in the accumulation of fluorine by deciduous plants. In species without any damage the control content of fluorine in leaves ranged from a 4-fold increase (ramanas rose) to a 16-fold increase (red elder). The highest volume of fluorine absorption was found in black cotoneaster, fragrant

Table 26. Deciduous trees and bushes suitable for planting in protection zones

Siberian pea shrub (*Caragana arborescens* Lam.)	+[a]
Flat-leaved birch (*Betula platyphylla* Sukacz.)	++
Red elder (*Sambucus racemosa* L.)	++++
Siberian elm (*Ulmus pumila* L.)	++
White dogwood (*Cornus alba* L.)	++++
Tartar honeysuckle (*Lonicera tatarica* L.)	++++
Bushy alder (*Duschekia fruticosa* (Rupr.) Pouzar.)	++
Goat willow (*Salix caprea* L.)	+
Common buckthorn (*Rhamnus cathartica* L.)	++
Black cotoneaster (*Cotoneaster melanocarpus* Lodd.)	++++
Ash-leaved maple (*Acer negundo* L.)	++
Alder buckthorn (*Frangula alnus* Mill.)	++
Silver oleaster (*Elaeagnus argentea* Pursh.)	++
Aspen (*Populus tremula* L.)	+++
Ramanas rose (*Rosa rugosa* Thunb.)	++
Acicular rose (*R. acicularis* Lindley)	+
Dove-coloured rose (*R. glauca* Pourr.)	+
Mountain ash-leaved sorbaria (*Sorbaria sorbifolia* (L.) A. Br.)	+++
Lilac (*Syringa vulgaris* L.)	+++
Mountain currant (*Ribes alpinum* L.)	+++
Oriental spiraea (*Spirea media* Fr. Schmidt)	+
Laurel-leaved poplar (*Populus laurifolia* Ldb.)	++++
Fragrant poplar (*P. suaveolens* Fischer)	+++
Siberian apple (*Malus pallasiana* Juz.)	+++

[a] ++++, The most resistant; +++, resistant; ++, moderately resistant; +, insufficiently resistant but with a high reducing ability.

poplar, bushy alder, red elder, oriental spiraea, aspen and alder buckthorn; these plants belonged to different groups in terms of the degree of damage. Thus, no relationship was found between the level of fluorine accumulation and the degree of damage in the deciduous plants tested. In conifers such a relationship is distinctly displayed, as has been shown.

Among the "accumulators" of fluorine of particular interest are plants which are able to combine a high fluorine content in the leaves and a low extent of leaf necrosis. Such combination is best demonstrated in black cotoneaster. Some species moderately or even severely damaged by fumigation in chambers demonstrated a good recovery in the next year: oriental spiraea, acicular rose, Siberian pea shrub and goat willow. They could be used under conditions of low and fairly moderate concentrations of fluorine. The plants of the most resistant and resistant groups can be used in the zone of high fluorine concentration and species such as laurel-leaved poplar, red elder and black cotoneaster can be planted in the immediate territory of the smelter.

Conclusion

Fluorine holds a particular position among industrial pollutants with regard to its toxicity and ability to exert a cumulative influence. The investigations presented are only a first step in studying its phytotoxicity. In our opinion further attention should be paid, first, to a comprehensive study of the effect of fluorine on coniferous forest ecosystems and second, to the elucidation of physiological-biochemical mechanisms in plants that come into action under the effect of fluorine. It is necessary to design a physiological model of the woody plant damaged by fluorine.

In order to preserve Siberian nature, especially in regions with a prevalence of broken ground, with low winter temperatures that facilitate the decline of trees weakened by pollutants and with abundant coniferous forests that have a low resistance to fluorine, the capacity of aluminium plants should be limited to 1000 tons of fluorine emitted into the atmosphere annually, with a further reduction if necessary. To construct smelters it is necessary, if possible, to choose open country sites. It should be remembered that a long time is required for the formation of woody plantations and it is necessary first of all to design protection zones and then not pay them less attention than the construction of the actual smelter. A search for and testing of forms of woody plants, including conifers, that are resistant to fluorine should be imperative.

There is also a need to establish a centre for phytotoxic investigations and to assess MPCs of toxic compounds from industrial emissions. This centre could join with the Institute of Plant Physiology to set up well-equipped stations and control centres in various regions of the country, including minimally polluted areas.

References

Abramashvili GG (1957) The impact of air pollution on coniferous forests. Hyg Sanit 4:67–69 (in Russian)

Adams DF, Hendrix JW, Applegate HG (1957) Atmospheric pollution: Relationship among exposure periods, foliar burn, and fluorine content of plants exposed to hydrogen fluoride. Agric Food Chem 5:108–116

Aleksandrov VA (1975) The cell, macromolecules and temperature. Nauka, Leningrad

Altergot VF (1966) Plant adaptation to elevated temperature of environment. In: Altergot VF (ed) Physiology of plant adaptation and resistance to unfavourable conditions. Nauka, Moscow, pp 17–25

Altergot VF (1969) Plant adaptation to elevated temperature of environment In: Altergot VF (ed) Physiology of plant adaptation and resistance during introduction. Nauka, Novosibirsk, pp 169–186

Altergot VF, Bukholtsev AN (1970) Types of defence responses of growing plants to cold. Physiol Biochem Cult Plants 2:148–153 (in Russian)

Amschl B (1986) Hauptthema: Forstschädliche Luftverunreinigung. Wald Holzwirt 34:14–15

Anisimova OA (1980) Formation of ecological complexes in forests weakened by fluorine emissions of aluminium plants. In: Isaev AS (ed) The role of dendrophilous insects in taiga ecosystems. Forest and Wood Institute, Krasnoyarsk, pp 5–6

Anisimova OA (1981) Xylophagous insects in pine forests exposed to fluorine emissions of the Irkutsk aluminium smelter In: Kulik SA (ed) Fauna and ecology of terrestrial arthropods of Siberia. Irkutsk University, Irkutsk, pp 102–111

Anisimova OA (1985) The effect of fluorine emissions of aluminium smelters on composition, ecology and number of coniferous xylophages. In: Isaev AS (ed) The monitoring system in forest protection. Forest and Wood Institute, Krasnoyarsk, pp 151–152

Anisimova OA, Sokov MK (1975) The role of insects in stands weakened by toxic emissions of aluminium smelters. In: Rozhkov AS (ed) The impact of anthropogenic and natural factors on conifers. Siberian Institute of Plant Physiology and Biochemistry, Irkutsk, pp 61–84

Antipov VG (1975) Pine species resistance to industrial gases. In: Yurkevich ID (ed) Botany. Research works, vol 17. Nauka i tekhnika, Minsk, pp 215–221

Antipov VG (1976) Spruce species resistance to industrial gases. In: Yurkevich ID (ed) Botany. Research works, vol 18. Nauka i tekhnika, Minsk, pp 204–208

Antipov VG (1978) The resistance of species of pseudotsuga, hemlock and larch to industrial gases. In: Yurkevich ID (ed) Botany. Research works, vol 20. Nauka i tekhnika, Minsk, pp 224–228

Antipov VG, Bolotov NA (1977) The response of fir species to air industrial pollution. In: Treshhevsky IV (ed) Protecting afforestation and forest cultures, vol 4. Voronezh Univ, Voronezh, pp 15–21

Applegate HG, Adams DF, Carriker RC (1960) Effect of aqueous fluoride solution on respiration of intact bush bean seedlings. I. Inhibition and stimulation of oxygen uptake. Am J Bot 47:339–345

Bach W (1985) "Waldsterben": our dying forests. Part III. Forest dieback: extent of damages and control strategies. Experientia 41:1095–1104

Baumeister W, Burghardt H (1957) Die Bedeutung der Elemente Zink und Fluor für das Pflanzenwachstum. Forschungsber Wirtsch Verkehrsministeriums Nordrhein – Westfalen No 388 (Quoted by Guderian 1979)

Belozerova LS (1964) The effect of sodium fluoride on photosynthesis, respiration and organic acid transformations in succulent plants in light. Trans Leningrad Univ Biol 4:116–121

Benedict HM, Ross JM, Wade RW (1964) The disposition of atmospheric fluorides by vegetation. Int J Air Water Pollut 8:279–289

Benedict HM, Ross JM, Wade RW (1965) Some responses of vegetation to atmospheric fluorides. J Air Pollut Control Assoc 15:253–255

Bengtsson G (1986) Inventory of forest damage in Sweden. Rapp Ingenjorsvetenskapsakad 310:17–25

Bennett JH, Hill AC (1974) Acute inhibition of apparent photosynthesis by phytotoxic air pollutants. In: Air pollution effects on plant growth. ACS Symp Ser 3, Washington, pp 115–127

Berenshtein FYa (1953) On the biological role of fluorine. Achievements of Present-Day Biol 35:216–228 (in Russian)

Berezhnykh ED (1978) Xylophagous insects of the damaged coniferous forests of the Baikal-Amur middle taiga zone. In: Rozhkov AS (ed) Conifers and dendrophagous insects. Siberian Institute of Plant Physiology and Biochemistry, Irkutsk, pp 37–51

Berry CR (1973) The differential sensitivity of eastern White pine to three types of air pollution. Can J For Res 3:543–547

Bodrov VA (1940) Forest melioration. Goslestekhizdat, Moscow

Bonneau M (1986) Forest decline in the East of France. Sci Sol 1:69–70

Bortitz S (1981) The impact of toxic compounds on plants. Lesnaya promyshlennost, Moscow, pp 47–73 (Translated from German)

Bowen HJM (1966) Trace elements in biochemistry. Academic Press, New York

Boyarkin AN (1951) A rapid method of determination of the peroxidase activity. Biochemistry 16:352–357 (in Russian)

Boyer JN, Houston DB, Jensen KF (1986) Impact of chronic SO_2, O_3 and SO_2 + O_3 exposures on photosynthesis of *Pinus strobus* clones. Eur J For Pathol 16:293–299

Braun G (1977) Über die Ursachen und Kriterien der Immissionsresistenz bei Fichten, *Picea abies* (L) Karst. II Reflektorische Immissionsresistenz. Eur J For Pathol 7:129–152

Brennan E, Rhoads AF (1976) The response of woody species to air pollutants in an urban environment. J Arboric 2:1–5

Butt VS (1980) Direct oxidases and related enzymes. Biochem Plants 2:81–123

Chang CW (1968) Effect of fluoride on nucleotides and ribonucleic acid in germinating corn seedling roots. Plant Physiol 43:669–674

Chang CW, Thompson CR (1956) Subcellular distribution of fluoride in navel orange leaves. Int J Air Water Pollut 9:685–691

Chuvaev PP, Kulagin YuZ, Getko NV (1973) Problems of industrial ecology and plant physiology. Nauka i tekhnika, Minsk

Cicák A (1979) Fluórointoxikacia a jej zakladné symptomy ataksie nepriameho vplyvu na topoloch. Folia Dendrol 5:89–108

Dässler H-G, Ranft H, Rehn K-H (1972) Zur Widerstandsfähigkeit von Gehölzen gegenüber Fluorverbindungen und Schwefeldioxid. Flora 161:289–302

De Cormis L (1970) Zanieczyszcenie atmosfery a roślinność. Ochr Pow 5:10–13

Dochinger LS (1973) Trees for polluted air. US Dep Agric For Serv Misc Publ No 1230 (Quoted by Kluczyński 1976)

Duda S (1981) Zawartosc wybranych metali ciezkich w igliwiu sosn y pospolitej (*Pinus Silvestris*) w rejonie oddziaywania emisji huty miedzi. Sylwan 125:119–127

Durmishidze SV, Ugrekhelidze DSh, Mitaishvili TI (1981) Metabolism of chemical pollutants of biosphere in plants. In: Slepyan GI (ed) The problems of phytohygiene and environmental protection. Zoological Institute, Leningrad, pp 112–117
Ennan AA, Takhtarova NS (1978) Ecological and economical problems of collecting and utilization of fluorine compounds from waste gases of aluminium and superphosphate production. In: Shukhardin SV (ed) Scientific and engineering revolution and the solution of ecological problems. Nauka, Moscow, pp 223–227
Ermakov AI (ed) (1972) Methods of biochemical analysis of plants. Kolos, Moscow
Ewert E, Lux H, Stein G, Enderlein H (1981) Forest management. In: Dässler H-G (ed) The effect of air pollution on vegetation. Lesnaya Promyshlennost, Moscow, pp 100–121 (Translated from German)
Fabijanowski J (1986) Hodowla lasu wobec zagrożenia imisjami drzewostanów w górach. Sylwan 130:53–66
Fink S (1986) Forest damage – a status report on effect via the leaves. Rapp Ingenjorsvetenskapsakad 310:27
Franke W, Heber U (1964) Über die quantitative Verteilung der Ascorbinsäure innerhalb der Pflanzenzelle. Z Naturforsch 169:1146–1149
Freedman B, Hutchinson TC (1980) Long-term effects of smelter pollution at Sudbury, Ontario, on forest community composition. Can J Bot 58:2123–2140
Garber K (1967) Luftverunreinigung und ihre Wirkungen. Gebrüder Borntraeger, Berlin
Gautier C, Clausmann R (1919) Ac Sci 112:105 (Quoted by Berenshtein 1953)
Genkel PA (1982) Physiology of heat and drought resistance of plants. Nauka, Moscow
Getko NV, Kulagin YuZ, Yafaev EM (1978) On gas-absorbing capacity of conifers. In: Kulagin YuZ (ed) Conifer ecology. Bashkiriya Branch USSR Acad Sci, Ufa, pp 112–120
Girs GI (1982) The physiology of the weakened tree. Nauka, Nobosibirsk
Glattes F (1985) Schwermetall und Ozongehalte in Waldökosystemen. Allg Forstztg 96:268–270
Godbold DL, Schlegel HI, Huttermann A (1985) A toxicity of heavy metals to spruce seedlings. In: Heavy metals and environment, vol 1. Int Conf, Athens. Sept 1985, Edinburgh, pp 500–502
Greshta Ya (1970) The effect of industrial air pollution on pine and fir stands. In: Kolesnikov BP (ed) Vegetation and industrial pollution. Ural Branch USSR Acad Sci, Sverdlovsk, pp 20–25
Greszta I, Braniewski St, Nosek A, Suchanek R (1981) Grenzwerte von aus verschiedenen Emissionsquellen stammenden Stäuben für die Entwicklung ausgewählter Arten von Waldbäumen. Mitt Forstl Bundes-Versuchsanst 137/2:199–209
Güde H (1981) Die Korrelationsrechnung als Beweisunterstützung und Schadensberechnungsbasis bei Rauchschäden im Wald. Mitt Forstl Bundes-Versuchsanst 137/1:87–91
Guderian R (1979) Air pollution, Ilkun GM (ed) Mir, Moscow, 200p (Translated from English)
Haagen-Smit AJ, Darley EF, Zaitlin M, Hull H, Noble W (1952) Investigation on injury to plants from air pollution in the Los Angeles area. Plant Physiol 27:18–34
Hager H, Kazda M (1985) The influence of stand density and canopy position on sulfur content in needles of Norway spruce (*Picea abies* L.). Water Air Soil Pollut 25:321–329
Härtel O, Fuchshofer H (1987) Neues zur Bioindikation von Abgaswirkungen auf Koniferen mittels des Trübungstest. Physiol Grundlagen und Methodisches. VDI-Ber 609:432–442
Hetz J, Richter J (1985) Immissionsschäden in Wäldern – eine Bibliographie. Hercynia 22:432–439
Hill AC, Pack MR, Transtrum LG, Winters WS (1959) Effects of atmospheric fluorides and various types of injury on the respiration of leaf tissues. Plant Physiol 34:11–16
Hitchcock AE, Zimmermann PW, Coe RR (1963) The effects of fluorides on milo maize. Contrib Boyce Thompson Inst 22:175–206
Horntvedt R, Robak H (1975) Relative susceptibility of eleven conifer species to fluoride air pollution. Medd Nor Inst Skogforsk 32:189–206

Hüttermann A, Ulrich B (1984) Solid phase-solution-root interactions in soils subjected to acid deposition. Philos Trans R Soc Lond 305:353–363

Ignatenko AA (1981) The effect of industrial pollution on the content of amino acids in resistant and sensitive plant species. Thesis, Kazan University

Ignatenko AA, Tarabrin VP (1975) Amino acid metabolism in woody plants under industrial pollution. In: Ponomarev AN (ed) Gas resistance of plants, vol 3. Perm Univ, Perm, pp 48–58

Ilkun GM (1971a) Gas resistance of plants. Naukova Dumka, Kiev

Ilkun GM (1971b) The effect of toxic gases on plants. Physiol Biochem Cult Plants 3:87–92 (in Russian)

Ilkun GM (1978) Air pollutants and plants. Naukova Dumka, Kiev

Ilkun GM, Bukolova TP (1974) Changes in pine needle structure under the impact of fluorine and chlorine. Ukr Bot J 31:624–629 (in Russian)

Ilkun GM, Motruk VV (1971) Restoration of cation–anion balance in leaves entered by ambient chlorine. II. The role of mineral ions. In: Ponomarev AN (ed) Gas resistance of plants, vol 2. Perm Univ, Perm pp 151–157

Ilkun GM, Motruk VV (1976) Accumulation and transport of fluoride compounds in soils. In: Ilkun GM (ed) Plants and industrial environment. Naukova Dumka, Kiev, pp 82–85

Ilkun GM, Motruk VV (1980) Absorption and removal of ions by plant roots in the polluted atmosphere. Plant Physiol 27:150–156 (in Russian)

Ilkun GM, Motruk VV, Mironova AS (1971) Restoration of cation–anion balance in leaves entered by sublethal doses of toxic gases. 1. The role of organic acids. In: Ponomarev AN (ed) Gas resistance of plants, vol 2. Perm Univ, Perm, pp 143–150

Ilkun GM, Yudin YuN, Kustovsky SE (1983) Sorption of polar gases by leaves of woody plants. Physiol Biochem Cult Plants 15:482–487 (in Russian)

Ishikawa N, Kobayashi E (1982) Flourine: Chemistry and application. Translated from Japanese. Mir, Moscow

IUFRO-Tagung in Marianske Lasne, 14–18 October 1974. Quoted by W Knabe, Luftverunreinigung und Forstwirtschaft, Forstarchiv (1975) 46:59–62

Ivanov VP (1973) Plant exudates and their importance for phytocenosis performance. Nauka, Moscow

Ivanova TM, Rubin BA (1962) On the nature of the phenoloxidase action of peroxidase. Biochemistry 27:622–630 (in Russian)

Jacobson JS, Weinstein LH. McCune DC, Hitchcock AE (1966) The accumulation of fluoride by plants. J Air Pollut Control Assoc 16:412–417

James GI, Courtin GM (1985) Stand structure and growth form of the birch transition community in an industrially damaged ecosystem, Sudbury, Ontario. Can J For Res 15:809–817

Jamrich V (1976) Some problems of plant resistance classification under conditions influenced by industrial air pollution. Zb ved pr Lesn fak vys skalesn drevarsk Zbolene 18:95–105

Jamrich V, Tomanova S, Kmet J (1989) Metabolicky stres a poruchy cinnosti listov vrby babylonsej (*Salix babylonica* L.) v podmienkach priemyselne fluorointoxikacie. Acta Fac For Zvolen 31:29–49

JAPCA (1971) Environmental protection agency sets national air quality standards. J Air Pollut Control Assoc 21:352–353

Jurat R, Schaub H, Stienen H, Bauch J (1986) Einfluss von Schwefeldioxid auf Fichten (*Picea abies* L. Karst.) in verschiedenen Bodensubstraten. Forstwiss Centralbl 105: 105–115

Kahle H, Bertela Ch, Noack G, Roder U, Ruther P, Breckle S-W (1989) Wirkung von Blei und Cadmium auf Wachstum und Mineralstoffhaushalt von Buchenjungwuchs. Allg Forstztg 44:786–788

Karolewski P, Bialobok S (1979) Wplyw dwutlenku siarki, ozonu, mieszaniny tych gasów i fluowodoru na uszkodzenie igiel modrzewia europejskiego. Arbor Kórnickie 24: 297–305

Kawecka A (1973) Changes of the leaves of the Scots pine (*Pinus silvestris* L.) due to the pollution of air with nitrogen compounds. Ecol Pol 21:105–120
Kazantseva EN (1966) Resistance of some cereals to fluorine. Vegetation and industrial pollution. Ural Branch USSR Acad Sci, Sverdlovsk, pp 45–47
Keller Th (1973) Zur Phytotoxizität staubförmiger Fluor-Verbindungen. Staub Reinhalt Luft 33:395–397
Keller Th (1974a) Translocation of fluoride in woody plants. Fluoride 7:31–35
Keller Th (1974b) The use of peroxidase activity for monitoring and mapping air pollution areas. Eur J For Pathol 4:11–19
Keller Th (1975a) Zur Phytotoxizität von Fluorimmissionen für Holzarten. Mitt Eidgenoss Anst Forstl Versuchsw 51:303–331
Keller Th (1975b) Zur Fluor-Translokation bei Waldbäumen. Mitt Eidgenoss Anst Forstl Versuchsw 51:337–358
Keller Th (1976) Zur Beeinflussung des physiologischen Blattalters von Waldbäumen durch Fluor-Immissionen. Mitt Eidgenoss Anst Forstl Versuchsw 52:101–158
Keller Th (1977) Begriff und Bedeutung der "latenten Immissionsschädigung". Allg Forst Jagdztg 148:115–120
Keller Th (1981) Die Beeinflussung physiologischer Prozesse der Fichte durch eine Winterbegasung mit SO_2. Mitt Forstl Bundes-Versuchsanst 137/1:115–120
Keller Th (1984) The influence of SO_2 on CO_2 uptake and peroxidase activity. Eur J For Pathol 14:354–359
Keller Th, Schwager H (1977) Air pollution and ascorbic acid. Eur J For Pathol 7:338–350
Kiennen L, Schuck HJ (1983) Untersuchungen über die Zuwachsentwicklung bei erkrankten Tannen. Eur J For Pathol 13:289–295
Kiseleva EK (1966) The analysis of fluorine containing compounds. Khimiya, Moscow
Kisser J (1964) Praktische und experimentelle Erfahrungen über die Fluor-Resistenz von Holzgewächsen. Vortrag Arbeitstagung "Fluor-Wirkung", Düsseldorf, März (Quoted by Garber 1967)
Klaus S (1989) Schwefel- und Nahrstoffgehalte in Pflanzenproben des österreichischen Bioindikatornetzes. In: Air pollution and forest decline: Proc 14th Int Meet Spec Air Pollut Effect Forests Ecosyst, Birmensdorf 2–8 Oct, 1988 vol 1, Int Union Forest Res Organ, pp 99–104
Klock GO (1980) Preplant soil fumigant increases survival and growth of field planted conifer seedlings. For Sci 26:400–402
Kluczyński B (1975) Wplyw zwiazków fluoru na stan zdrowotny drzew i krzewów przy Hucie Aluminium "Konin". Arbor Kórnickie 20:317–343
Kluczyński B (1976) Oddzialywanie fluoru i jego swiazków na rośliny. Arbor Kórnickie 21:401–418
Knabe W (1975) Luftverunreinigungen und Forstwirtschaft. Forstarchiv 46:59–62
Knabe W, Pohlmann H, Urber W (1989) Statistische Überprüfung der Silizium-Hypothese in der Waldschadensforschung. Forstarchiv 60:223–227
Knunyants IL, Fokin AV (1963) The taming of an unassailable element. USSR Acad Sci, Moscow
Kovalev IE, Polevaya OYu (1981) Antibodies for physiologically active compounds. Meditsina, Moscow
Kozyukina ZhT (1976) On some ecologo-physiological parameters of gas resistance of woody plants. Sci Rep Higher School (Biol Sci) 7:103–107
Kramer PD, Kozlovsky TT (1983) Physiology of woody plants. Translated from English. Lesnaya Promyshlennost, Moscow
Krasinsky NP (1940) On physiological nature of gas resistance of plants. Proc Gorky Univ 9:3–74
Krasinsky NP (1950) Theoretical fundamentals for ranking plant species in terms of gas resistance. In: Krasinsky NP (ed) Smoke resistance of plants and smoke resistant groups. Gorky, Moscow, pp 9–113
Kretovich VL (1980) Plant Biochemistry. Vysshaya shkola, Moscow

Kronberger W, Halbwachs G, Richter H (1978) Fluortranslokation in *Picea abies* (L) Karsten. Angew Bot 52:149–154

Krouse HR, Legge AH, Brown HM (1984) Sulphur gas emissions in the boreal forest: the West Whitecourt case study. Stable sulphur isotopes. Water Air Soil Pollut 22:321–347

Kulagin YuZ (1974) Woody plants and industrial environment. Nauka, Moscow

Kulagin YuZ (1980) Forest forming species, industrial impact and prognosis. Nauka, Moscow

Kulagin YuZ (1985) Industrial dendroecology and prognosis. Nauka, Moscow

Landa M (1986) Immissionen und Waldschäden in der Tschechoslowakei. Allg Forstztg 97:210

Landmann G, Bonneau M, Adrian M (1987) Le deperissement du sapin pectine et de l'epicea commun dans le massif Vosgien est-il en relation avec l'etat nutritionnel des peuplements? Rev For Fr 1:5–11

Landolt W, Keller Th (1985) Uptake and effects of air pollutants on woody plants. Experientia 41:301–310

Landolt W, Scherrer HU (1985) Immissionskündliche Untersuchung der Waldschäden im Unterwallis. Schweiz Z Forstwes 136:733–747

Latocha E (1983) Próby ustalenia progu toksyczności dwutlenku siarki dla drzewostanów iglastych. Sylwan 127:21–27

Lenzburg N (1985) Erfassung der Immissionsschadensentwicklung in einer Forstverwaltung. Schweiz Z Forstwes 136:369–382

Liebold E, Ranft H (1981) The practice of plant adaptation in forestry. In: Dässler HG (ed) The effect of air pollution on vegetation. Lesnaya promyshlennost, Moscow, pp 139–158 (Translated from German)

Lines R (1981) Performance of different species and seed origin in the industrial Pennines of northern Britain. Arch Ochr Srodowiska 2/4:107–120

Ling-Ling W, Miller GW (1972) Effect of HF on the fine structure of mesophyll cell from *Glycine max*. Merr. Fluoride Q Rep 5:67-73

Löffler A (1983) Regressionsanalytische Bewertung der Auswirkungen von Exhalaten auf die Reaktion der Waldböden in dem Immissionsgebiet von Ziar nad Hronom. Acta Inst For Zvolenensis 6:233–249

Lomagin AG (1969) The effect of light on plant cell resistance to injury. Achievements of Present-Day Biol 67:147–163 (in Russian)

Lyanguzova IV, Chertov OG (1990) Chemical composition of plants under air and soil pollution. In: Alekseev VA (ed) Forest ecosystems and air pollution. Nauka, Moscow, pp 75–86

Magel E, Ziegler H (1986) Einfluβ von Mineralstoffernährung, Ozon und saurem Nebel auf den Gehalt an Adeninnukleotiden, anorganischem Phosphat und Kohlenhydraten in Nadeln von *Picea abies* (L.) Karst. Forstwiss Centralbl 105:243–251

Mamaev SA, Makarov NM (1976) Individual variability of common pine regarding resistance to sulphur dioxide. In: Ilkun GM (ed) Plants and industrial environment. Naukova Dumka, Kiev, pp 96–99

Materna J (1975) Les a prumysl. Lesnictví 21:211–220

Materna J (1978) Lesní hospodářství a zemědělství v oblastech postižených imisemi. Pr Vyzk Ustavu Lesn Hospod Myslivosti (Strnady) 53:99–115

Mathy P (1977) Forests et polution atmosphérique. Bull Soc R For Belg 84:226–233

McCune DC, Weinstein LH (1971) Metabolic effects of atmospheric fluoride on plants. Environ Pollut 1:169–174

McCune DC, Weinstein LH, Jacobson JS, Hitchcock AE (1964) Some effects of atmospheric fluoride on plant metabolism. J Air Pollut Control Assoc 14:465–468

McNulty IB, Newman DW (1956) The effect of a lime spray on the respiration rate and chlorophyll content of leaves exposed to atmospheric fluoride. Proc Utah Acad Sci 33:73–79

McNulty IB, Newman DW (1957) Effect of atmospheric fluoride on the respiration rate of bush bean and gladiolus leaves. Plant Physiol 32:121–124

McNulty IB, Newman DW (1961) Mechanism of fluoride induced chlorosis. Plant Physiol 36:385–388

Mejnartowicz LE, Zukasiak H (1985) Level of sugars in Scots pine trees of different sensitivity to fluoride and sulphur dioxide. Eur J For Pathol 15:193–198
Menser HA, Heggestand HE (1966) Ozone and sulfur dioxide synergism injury to tobacco plants. Science 153:424–425
Mielke U (1977) Die Wirkung von Luftverunreinigungen auf Pflanzen. Hercynia 14: 84–100
Mikhailova TA, Anisimova OA (1978) The effect of fluorine on long-horn beetles of the genus *Monochamus*. In: Rozhkov AS (ed) Conifers and dendrophilous insects. Siberian Institute of Plant Physiology and Biochemstry, Irkutsk, p 126
Miller HG (1984) Deposition-plant-soil interactions. Philos Trans R Soc 305:339–351
Molchanov AA (1961) Forest and climate. USSR Acad Sci, Moscow
Morel C, Chaouard P (1967) Etude comparative de l'acion de l'acide fluorhydride et de l'emhydride sulfureux sur le sarras in *Poligonum fagopurum* L. Oecol Plant 2:27–42
Mowszowicz J (1980) Dendrologia a przemysl. Rocz Sekc Dendrol Pol Tow Bot 33: 137–146
Negrutskaya GM (1968) Variation in the content of sulphur and nitrogenous compounds in pine needles damaged by toxic gases. In: Ilkun GM (ed) Plant and industrial environment. Naukova Dumka, Kiev, pp 105–111
Negrutskaya GM (1970) Physiologo-biochemical processes in pine affected by toxic gases. Thesis, Kiev University
Negrutskaya GM, Popov VA, Gunko AV, Sandakova TV (1981) The effect of sulphur dioxide on some physiological and morphological parameters of common pine. In: Nikolaevsky VS (ed) Physiologo-biolchemical mechanisms of damage and plant resistance. Nauka, Novosibirsk, pp 126–131
Negrutsky SF, Popov VA, Prisedsky YuG, Eremka EV (1981) Mechanisms of penetration of pollutants into the higher plant organism. In: Slepyan EI (ed) Problems of phytohygiene and protection of environment. Zoological Institute, Leningrad, pp 104–108
Nikitin DP, Novikov YuV (1986) Environment and man. Vysshaya shkola, Moscow
Nikolaevsky VS (1965) Some aspects of oxidation-reduction processes in woody plants in terms of their gas resistance. Proc Biol Inst Ural Branch USSR Acad Sci 43: 139–146
Nikolaevsky VS (1968) The role of some oxidation systems in respiration and gas resistance of plants. Plant Physiol 15:110–115 (in Russian)
Nikolaevsky VS (1979) Biological principles of gas resistance of plants. Nauka, Novosibirsk
Nikolaevsky VS (ed) (1984) Provisional standards of maximum permissible concentrations of air pollutants adversely affecting forests in the vicinity of the museum "Yasnaya polyana". Central Library of Scientific and Technical Information, Moscow
Nikolaevsky VS, Miroshnikova AT (1974) Permissible standards of air pollution for plants. Hyg Sanit 4:l6–18 (in Russian)
Nikolaevsky VS, Pershina NA (1981) The problem of maximum permissible concentrations of pollutants affecting plants and their communities. In: Slepyan EI (ed) Problems of phytohygiene and environmental pollution. Zoological Institute, Leningrad, pp 117–121
Noack K (1920) Untersuchungen über lichtkatalysche Vorgänge von physiologischer Bedeutung. Z Bot 12:273–347
Noack K (1925) Photochemische Wirkungen des Chlorophylls und ihre Bedeutung für die Kohlensäureassimilation. Z Bot 17:481–548
Novitskaya YuE, Chikina PF (1980) Nitrogen metabolism of pine in the North. Nauka, Leningrad
Obraztsova VI, Kozyukina ZhT (1971) Some specific features of nitrogen metabolism under air pollution. In: Ilkun GM (ed) Plants and industrial environment. Naukova Dumka, Kiev, pp 61–63
Peters RA, Shorthouse M (1964) Fluoride metabolism in plants. Nature 202:21–24
Petrovskaya-Baranova TP (1983) Physiology of adaptation and plant introduction. Nauka, Moscow

Pindyura ES, Pleshanov AS (1975) Morphometric responses of conifers on defoliation. In: Rozhkov AS (ed) The impact of anthropogenic and natural factors on conifers. Siberian Inst Plant Physiol Biochem, Irkutsk, pp 159–178

Pindyura ES, Stepanchuk ES (1978) Variations in early/late wood ratio in conifers in the experiment. In: Rozhkov AS (ed) Conifers and dendrophilous insects. Siberian Inst Plant Physiol Biochem, Irkutsk, pp 105–114

Pleshanov AS, Pindyura ES (1975) Pathology of the Siberian cedar pine in outbreaks of the Siberian pine moth In: Rozhkov AS (ed) The impact of anthropogenic and natural factors on conifers. Siberian Inst Plant Physiol Biochem, Irkutsk, pp 215–240

Pleshkov BP (1976) Practical course on plant biochemistry. Kolos, Moscow

Plokhinsky NA (1970) Biometrics. Moscow Univ, Moscow

Popov VA, Negrutskaya GM, Shishmareva AT, Rozhkovsky AD (1981) Changes in the content of nitrogenous substances in leaves of woody plants exposed to nitrogen and ammonia oxides. In: Nikolaevsky VS (ed) Physiologo-biochemical mechanisms of damage and resistance of plants. Nauka, Novosibirsk, pp 116–126

Popovichev BG (1980) The effect of gases emitted by industrial enterprises on indices of the seed quality in common pine and white birch. Silvic For Cult Soil Sci 9:59–62

Popovichev BG (1987) Populations of xylophagous insects in industrial ecosystems. In: Soloviev VA (ed) Ecology and forest protection. Forest Engineering Acad, Leningrad, pp 10–13

Prise V (1976) Analytical atomic-absorption spectrophotometry. Mir, Moscow

Prokushkin SG (1982) Mineral nutrition of pine in cold soils. Nauka, Novosibirsk

Pylypes B, Redmann RE (1984) Acid-buffering capacity of foliage from boreal forest species. Can J Bot 62:2650–2653

Rachkovskaya MM, Kim LO (1980) Changes in the activity of some oxidases as an indicator of plant adaptation to industrial pollution. In: Nikolaevsky VS (ed) Gas resistance of plants. Nauka, Novosibirsk, pp 117–126

Ratner EI, Akimochkina TA (1949) The effect of cations on some intercellular processes in plants depending on the source of absorption of these cations by roots. News Acad Sci USSR Biol Ser 3:296–315

Raunemaa T, Hari P, Kukkonen J, Anttila P, Katainen H-S (1987) Long-term changes of needle litter in coniferous forests of Finland. Can J For Res 17:466–471

Reepmeyer H, Vetter H (1976) Der Einfluss von Fluorwasserstoff im Vergleich zu dem von Calciumfluoridstaub auf Omorica-Fichten. Ziegelindustrie 11:485–492

Reigber E, Brawn G (1985) Forstliche Bioindikator-untersuchungen in Bayern. Methodik und erste Ergebnisse 1981/82. Schriftenr Forstwiss Fak Univ München und Bayer Forstl Versuchs Forschungsanst 68:179

Rinkis RYa, Nollendorf VF (1982) Balanced nutrition of macro- and microelements. Zinatne, Riga

Robak H (1969) Some facts about the absorption of fluorine by 1 and 2 years old needles of conifers in the environment of aluminium plants. Tidsskr Skogbruk 77:1–22 (in Norwegian)

Rohmeder E, Schönborn AV (1965) Der Einfluss von Umwelt und Erbgut auf die Widerstandsfähigkeit der Waldbäume gegenüber Luftverunreinigung durch Industrieabgase. Ein Beitrag zur Züchtung einer relativ rauchresistenten Fichtensorte. Forstwiss Centralbl 84:1–68

Rozhkov AS (1981) The tree and the insect. Nauka, Novosibirsk

Rozhkov AS, Byalaya IV (1966) The trunk pests. In: Rozhkov AS (ed) The pests of the Siberian larch. Nauka, Moscow, pp 19–22

Rozhkov AS, Massel GI (1982) Resin substances of conifers and xylophagous insects. Nauka, Novosibirsk

Rozhkov AS, Vasilieva TG (1981) On physiological state of pine forests weakened by industrial emissions on the eastern coast of lake Baikal. In: Galazy GI (ed) Efficient use of natural resources of reservoirs and their protection. Limnological Inst, Irkutsk, pp 139–141

Rozhkov AS, Kholopova RV, Massel GI et al. (1975) Pathology and resistance of pine and larch in temporarily flooded low-lying forests in south-eastern Siberia. In: Rozhkov AS (ed) The effect of anthropogenic and natural factors on conifers. Siberian Inst Plant Physiol Biochem, Irkutsk, pp 95–140
Rubin BA, Artsikhovskaya EV (1960) Biochemistry and physiology of plant immunity. USSR Acad Sci, Moscow
Rubin BA, Ladygina ME (1974) Physiology and biochemistry of plant respiration. Nauka, Moscow
Rudolph E, Bukatsch F (1966) Die Protochlorophyll(id)-Umwandlung und ihre Beziehung zur Photooxidation der Ascorbinsäure bei etiolierten Keimpflanzen. Planta 69:124–134
Ryazanov VA (1949) Basic principles of public health standards for air pollution. Hyg Sanit 5:3–9 (in Russian)
Ryškova L (1982) Ovlivnění sazenic několika druhu smrku kyslicníkem siričitým. Pr Vyzk Ustavu Lesn Hospod Myslivosti (Strnady) 60:85–97
Sabashnikov V (1911a) The effect of tar vapours on vegetation. Plant Dis 1/2:1–7
Sabashnikov V (1911b) The influence of coal fumes on ambient vegetation. Plant Dis 1/2:42–82
Sadilova MC (1964) Industrial emissions of the cryolite plant and their effect on environment. In: Kolesnikov BP (ed) Nature protection in the Ural, vol 4. Ural Branch USSR Acad Sci, Sverdlovsk, pp 43–48
Saimons H (1953) Hydrogen fluoride, vol 1. Translation from English. Inostrannaya Literatura, Moscow, pp 192–220
Salyaev RK (1969) The uptake of substances by the plant cell. Nauka, Moscow
Selikhovkin AV (1982) The effect of hydrogen fluoride on development of the gypsy moth. In: Soloviev VA (ed) Ecology and forest protection. Forest Engineering Acad, Leningrad, pp 83–86
Selikhovkin AV (1985) The effect of industrial pollution on needle-leaf-biting insects. Thesis, Leningrad University
Selikhovkin AV (1987) The formation of specific entomocomplexes of phyllophages under conditions of intense industrial pollution. In: Soloviev VA (ed) Ecology and forest protection. Forest Engineering Acad, Leningrad, pp 5–10
Serebryakova LK (1980) Permissible concentrations of toxic air pollutants for arboreal vegetation. In: Nikolaevsky VS (ed) Gas resistance of plants. Nauka, Novosibirsk, pp 184–185
Shakhov AA (1956) Salt tolerance of plants. Nauka, Moscow
Shlyk AA (1965) Analyses of chlorophyll and carotenoids in green leaf extracts. In: Pavlinova OA (ed) Biochemical methods in plant physiology. Nauka, Moscow, pp 154–169
Sitnikova AS (1964) On the study of physiological parameters of woody and shrubbery species in terms of gas and smoke resistance. In: Kolesnikov BP (ed) Vegetation and industrial pollution. Ural Branch USSR Acad Sci, Sverdlovsk, pp 39–44
Skiba U, Peirson-Smith TJ, Cresser MS (1986) Effects of simulated precipitation acidified with sulphuric and/or nitric acid on the throughfall chemistry of sitka spruce *Picea sitchensis* and heather *Calluna vulgaris*. Environ Pollut 11:255–270
Skuodene L (1984) The effect of industrial emissions on some physiological functions in woody species. In: Vajchis MV (ed) Abstr of All-Union Meet. Latviun Inst of Forestry, 26–27 June 1984, Kaunas, pp 57–58
Slater EC, Bonner WD (1952) The effect of fluoride on the succinic oxidase system. Biochem J 52:185–196
Smirnova VA (1968) Winter hardiness and frost resistance of woody plants. Nauka i tekhnika, Minsk
Smith UH (1985) The forest and the atmosphere. Progress, Moscow (Translated from English)
Sokov MK, Gabib-Zade LA (1977) Intraspecific differences in gas resistance of the pine. In: Salyaev RK (ed) Physiologo-biochemical and ecological aspects of plant resistance

to unfavourable environmental factors. Siberian Inst Plant Physiol Biochem, Irkutsk, pp 213–220
Sokov MK, Rozhkov AS (1975) Dynamics of increment reduction in conifers affected by industrial emissions of aluminium smelters. In: Rozhkov AS (ed) The effect of anthropogenic and natural factors on conifers. Siberian Inst Plant Physiol Biochem, Irkutsk, pp 9–60
Solberg RA, Adams DF (1956) Histological responses of some plant leaves to hydrogen fluoride and sulfur dioxide. Am J Bot 43:755–760
Spakhova AS, Ryazantseva LA (1978) The injury of some woody plants by sulphur dioxide. Plant Physiol 25:407–409 (in Russian)
Sterba H (1984) Theoretische Überlegungen zur ertragskundlichen Auswirkung von Waldschäden. Forstarchiv 55:34–36
Strogonov BL (1962) Physiological principles of salt tolerance of plants. Nauka, Moscow
Styrikovich MA, Vnukov AK (1981) Ecological problems of energetics. In: Izrael YuA (ed) Comprehensive analysis of environment Gidrometeoizdat, Leningrad, pp 70–77
Sudachkova NE (1964) Development of root systems and nitrogen-phosphorus metabolism in coniferous regrowth under mountain conditions. Proc Sib Branch USSR Acad Sci Biol Ser 4:22–27
Sudachkova NE (1977) Metabolism of conifers and wood formation. Nauka, Novosibirsk
Surtaev NKh (1985) The effect of industrial emissions on stands around Krasnoyarsk city. In: Belov SV (ed) Forest taxation and forest management. Krasnoyarsk Univ, Krasnoyarsk, pp 125–128
Swieboda M (1964) Niektóre zagadnienia wplywu na lasy przemyslowych zanieczyszczeń powietrza zwiazkami fluoru. Sylwan 6:45–54
Tarabrin VP (1974) The resistance of woody plants under industrial pollution of the environment. Thesis, Kiev University
Tarabrin VP, Bashkatov VG (1986) The combined effect of pyridine, phenol and sulphur dioxide on plants. In: Kondratyuk EN (ed) Phytotoxicity of organic and inorganic pollutants. Naukova Dumka, Kiev, pp 75–78
Tarabrin VP, Ignatenko AA (1976) Metabolism of free amino acids in leaves and roots of woody plants under industrial pollution. In: Ilkun GM (ed) Plants and industrial environment. Naukova Dumka, Kiev, pp 129–130
Tarabrin VP, Korshikov (1986) Ecological aspects of pollution and toxic effect of organic pollutants. In: Kondratyuk EN (ed) Phytotoxicity of organic and inorganic pollutants. Naukova Dumka, Kiev, pp 37–78
Tarchevsky VV (1964) The effect of smoke and gas emissions of Ural industrial enterprises on vegetation. In: Kolesnikov BP (ed) Plants and industrial environment. Ural Branch of the USSR Acad Sci, Sverdlovsk, pp 5–69
Tarchevsky IA (1977) Fundamentals of photosynthesis. Nauka, Moscow
Ten Houten JG (1974) Air pollution and horticulture. Proc XIX Int Horticult Congr, vol 2. Warszawa, 11–18 September 1979, PP 57–71
Teuchert E, Teuchert I (1983) Möglichkeiten der Anreicherung von Schwefelsäure in Baumkronen. Forstwiss Centralbl 102:181–186
Thiel WR (1985) Der Patient stirbt – die Ärzte diskutieren weiter. Umwelt 4:356–358
Timiryazev KA (1957) Selected works, vol 1. State Publ House for Agricultural Literature, Moscow, 723 pp
Tkachenko ME (1952) General silviculture. Goslesbumizdat, Moscow
Thomas MD (1962) The effect of air pollution on plants. In: Goldberg MS (ed) Air pollution. World Health Organization, Geneva. USSR Ministry of Public Health, Moscow, pp 263–306 (Translated from English)
Torgasheva EG (1984) Ultrastructural changes in chloroplasts of the leaf mesophyll in *Morus alba* (Moraceae) around the aluminium smelter. Bot J 69:921–925 (in Russian)
Treshow M, Pack MR (1970) Fluoride. In: Jacobson JS, Hill AC (ed) Recognition of air pollution injury to vegetation: a pictorial atlas. Air Pollution Control Association, Pittsburgh, Pennsylvania 1970 (Quoted by Horntvedt and Robak 1975), pp D1–D17

Tzschacksch O (1983) Möglichkeiten der Begründung von Fichtenbeständen verminderter Anfälligkeit gegenüber phytotoxischen Immissionen durch Hochpfropfungen. Soz Forstwirt 33:339–341

Udovenko GV (1973) The character of the defence adaptive response and reasons for different plant resistance to extreme impacts. In: Dorofeev VF (ed) Trans Applied Bot, Genetics and Selection, vol 49. All-Union Inst of Plant Growing, Leningrad, pp 258–268

Ulrich B (1980) Die Wälder in Mitteleuropa: Messergebnisse ihrer Gefährdung, Prognose ihrer Entwicklung. Allg Forstzeitschrift 35:1198–1202

USSR Ministry of Public Health (1979) Manual on air pollution control. Gidrometeoizdat, Leningrad

Vainshtein EA, Soldatenkov SV (1965) The effect of sodium fluoride on respiration of leaves of kidney beans and acid formation in them. Trans Leningrad Univ Biol 9:113–117

Valter OA, Pinevich LM, Varasova NN (1957) Manual on plant physiology with biochemistry essentials. Selkhozgiz, Moscow

Van Haut H (1975) Kurzzeitversuche zur Ermittlung der relativen Phytotoxizität von Stickstoffdioxid. Staub-Reinhalt 35:193–198

Vinogradov AP (1957) Geochemistry of rare and trace chemical elements in soils. Nauka, Moscow

Vlasyuk PA, Mitsko VN (1967) The effect of fluorides on the activity of some pea enzymes. In: Vlasyuk PA (ed) Microelements in agriculture and medicine, vol 3. Naukova dumka, Kiev, pp 46–49

Volkova LM, Vasilieva TG (1980) Specific biochemical composition of leaves of birch and its entomofauna in forests polluted by aluminium smelter emissions. In: Isaev AS (ed) The role of dendrophilous insects in taiga ecosystems. Forest and Wood Institute, Krasnoyarsk, pp 22–24

Vysotsky GN (1938) On hydrological and meteorological impacts of forests. Goslestekhizdat, Moscow

Walberg G (1976) Skadevirkningen av fluoridhoidig stov pa furu (*Pinus sylvestris*) ved aluminimverk. Tidsskr Skogbruk 84:271–286

Warburg O, Christian W (1942) Isolierung and Kristallisation des Gärungsferments Enolase. Biochem Z 310:384–421

Wehner O (1928) Untersuchungen über die chemische Beeinflussbarkeit des Assimilationsapparates. Planta 6:543–590

Weidenbach P (1984) Welche waldbaulichen Möglichkeiten ergeben sich in geschädigten Beständen? Allg Forstztg 39:758–761

Weiger H (1983) Der Wald stirbt. Prax Naturwiss Biol 32:278–287

Weinstein LH (1971) Boyce Thompson Institute for plant research. Jonkers, New York, Schreiben vom 15.1.1971 an den Verein Deutscher Ingenieure, Düsseldorf (Quoted by Guderian 1979)

Wentzel KF (1963) Waldbauliche Massnahmen gegen Immissionen. Allg Forstztg 18:101–106

Wentzel KF (1967) Vorschläge zur Klassifikation der Immissionserkrankungen. Forstarchiv 38:77–79

Wentzel KF (1968) Empfindlichkeit und Resistenzunterschiede der Pflanzen gegenüber Luftverunreinigung. Forstarchiv 39:189–195

Wentzel KF (1980) Immissionsshäden und Sozialpflichtigkeit der Forstwirtschaft. Forst und Holzwirt 35:409–410

Yager K (1975) Methods of control and regulation of air pollution. In: Izrael YuA (ed) Comprehensive analysis of environment. Gidrometeoizdat, Leningrad, pp 280–304

Yang SF, Miller GW (1963) Biochemical studies on the effect of fluoride on higher plants. I. Metabolism of carbohydrates, organic acids and amino acids. Biochem J 88:505–509

Yang Y-S, Skelly JM, Chevone BI (1982) Clonal response of eastern white pine to low doses of O_3, SO_2 and NO_2 singly and in combination. Can J For Res 12:803–808

Zimmerman PW, Hitchcock AE (1956) Susceptibility of plants to hydrofluoric acid and sulfur dioxide gases. Contrib Boyce Thompson Inst 18:263–279

Zwiazek JJ, Shay JM (1988a) The effect of sodium fluoride on cytoplasmic leakage and the lipid and fatty acid composition of jack pine (*Pinus banksiana*) seedlings. Can J Bot 66:535-541

Zwiazek JJ, Shay JM (1988b) Sodium fluoride induced metabolic changes in jack pine seedlings. II. Effect on growth, acid phosphatase, cytokinins, and pools of soluble proteins, amino acids, and organic acids. Can J For Res 18:1311–1317

Zwiazek JJ, Shay JM (1988c) Sodium fluoride induced metabolic changes in jack pine seedlings. I. Effect on gas exchange, water content, and carbohydrates. Can J For Res 18:1305–1310

Subject Index

MIX
Papier aus verantwortungsvollen Quellen
Paper from responsible sources
FSC® C105338

If you have any concerns about our products,
you can contact us on
ProductSafety@springernature.com

In case Publisher is established outside the EU,
the EU authorized representative is:
Springer Nature Customer Service Center GmbH
Europaplatz 3, 69115 Heidelberg, Germany

Printed by Libri Plureos GmbH
in Hamburg, Germany